Studies in Computational Intelligence

Volume 539

Series editor

Janusz Kacprzyk, Polish Academy of Sciences, Warsaw, Poland
e-mail: kacprzyk@ibspan.waw.pl

For further volumes:
http://www.springer.com/series/7092

About this Series

The series "Studies in Computational Intelligence" (SCI) publishes new developments and advances in the various areas of computational intelligence—quickly and with a high quality. The intent is to cover the theory, applications, and design methods of computational intelligence, as embedded in the fields of engineering, computer science, physics and life sciences, as well as the methodologies behind them. The series contains monographs, lecture notes and edited volumes in computational intelligence spanning the areas of neural networks, connectionist systems, genetic algorithms, evolutionary computation, artificial intelligence, cellular automata, self-organizing systems, soft computing, fuzzy systems, and hybrid intelligent systems. Of particular value to both the contributors and the readership are the short publication timeframe and the world-wide distribution, which enable both wide and rapid dissemination of research output.

Martine Ceberio · Vladik Kreinovich

Editors

Constraint Programming and Decision Making

Springer

Editors
Martine Ceberio
Department of Computer Science
University of Texas at El Paso
El Paso Texas
USA

Vladik Kreinovich
Department of Computer Science
University of Texas at El Paso
El Paso Texas
USA

ISSN 1860-949X
ISBN 978-3-319-38202-9
DOI 10.1007/978-3-319-04280-0
Springer Cham Heidelberg New York Dordrecht London

ISSN 1860-9503   (electronic)
ISBN 978-3-319-04280-0   (eBook)

Printed on acid-free paper

Springer is part of Springer Science+Business Media (www.springer.com)

Preface

Constraint programming and decision making are important. Constraint programming and decision-making techniques are essential in the building of intelligent systems. They constitute an efficient approach to representing and solving many practical problems. They have been applied successfully to a number of fields, such as scheduling of air traffic, software engineering, networks security, chemistry, and biology. However, despite the proved usefulness of these techniques, they are still under-utilized in real-life applications. One reason is the perceived lack of effective communication between constraint programming experts and domain practitioners about constraints, in general, and their use in decision making, in particular.

CoProd workshops. To bridge this gap, annual International Consstraint Porgramming and Decision Making workshops CoProd'XX have been organized since 2008: in El Paso, Texas (2008, 2009, 2011, and 2013), in Lyon, France (2010), and in Novosibirsk, Russia (2012); CoProd'2014 will be held in Würzburg, Germany. This volume contains extended version of selected papers presented at previous CoProd workshops.

CoProD workshops aim to bring together, from areas closely related to decision making, researchers who design solutions to decision-making problems and researchers who need these solutions and likely already use some solutions. Both communities are often not connected enough to allow cross-fertilization of ideas and practical applications.

CoProD workshops aim at facilitating networking opportunities and cross-fertilization of ideas between the approaches used in the different attending communities. Because of this, in addition to active researchers in decision making and constraint programming techniques, these workshops are also attended by domain scientists – whose participation and input is highly valued in these workshops.

The *goal of CoProD workshops is therefore to constitute a forum for inter-community building*. The objectives of this forum are to facilitate:

- The presentation of advances in constraint solving, optimization, decision making, and related topics;
- The development of a network of researchers interested in constraint techniques, in particular researchers and practitioners that use numeric and symbolic approaches (or a combination of them) to solve constraint and optimization problems;
- The gap bridging between the great capacity of the latest decision-making/constraint techniques and their limited use.

CoProD workshops can impact these communities by easing collaborations and therefore the emergence of new techniques, and by creating a network of interest.

The objectives of CoProD are also relayed all year round through the website constraintsolving.com.

Topics of interest. The main emphasis is on the joint application of constraint programming and decision making techniques to real-life problems. Other topics of interest include:

- Algorithms and applications of:
  - Constraint solving, including symbolic-numeric algorithms
  - Optimization, especially optimization under constraints (including multi-objective optimization)
  - Interval techniques in optimization and their interrelation with constraint techniques
  - Soft constraints
  - Decision making techniques
- Description of domain applications that:
  - Require new decision making and/or constraint techniques
  - Implement decision making and/or constraint techniques

Contents of the present volume: general overview. All these topics are represented in the papers forming the current volume. These papers cover all the stages of decision making under constraints:

- how to formulate the problem of decision making in precise terms, taking different criteria into account;
- how to check whether (and when) the corresponding decision problem is algorithmically solvable;
- once we know that the decision problem is, in principle, algorithmically solvable, how to find the corresponding algorithm, and how to make this algorithm as efficient as possible;
- how to take into account uncertainty, whether it is given in terms of bounds (intervals), probabilities, or fuzzy sets?

How to formulate the problem of decision making: general case. The paper [2] emphasizes that in decision making, it is important *not to oversimplify* the problem: a model which is a reasonably good (but not very accurate) fit for all previous observations can lead to misleading decisions. A similar conclusion is made in [11]: if we try to simply our problem by ignoring some of the natural constraints, often, the problem, instead of becoming algorithmically easier, becomes more complex to solve. So, using a more realistic model not only makes the results of the computations more adequate, it also often makes computations themselves easier (and faster).

Case of multi-criterion decision making. How can we combine different criteria?

- It is desirable to find a combination rule which is in best accordance with the actual decisions; a new method for solving the corresponding optimization problems is presented in [14].

– In many practical situations, it is possible to use known *symmetries* to find the most appropriate combination.
  • The paper [10] uses symmetries to explain why tensors and polynomial combination rules are often practically useful.
  • The paper [7] applies similar symmetry ideas to a specific problem of selecting the best location for a meteorological tower.

Instead of combining *criteria* and solving the resulting combined optimization problem, we can alternatively solve the optimization problems corresponding to all possible combinations, and then *select the solution* which is, in some reasonable sense, the most appropriate. As shown in [21], in this case also natural symmetries explain the efficiency of empirically successful selection heuristics.

When are problems algorithmically solvable?

– For *general* decision making problems, this question is analyzed in [1]; this paper also analyzes when it is possible to solve the problem while avoiding making irreversible changes.
– In some cases, when no algorithm is possible for a general *mathematical problem*, algorithms becomes possible is we only consider *physically meaningful* cases, i.e., if we take into account additional physical constraints [12].
– In some cases, partial solutions can be extended to general ones:
  • Paper [8] shows that, in principle, it is sufficient to be able to algorithmically compute the *quality* of the best decision, then it is possible to algorithmically find this optimal decision.
  • The paper [3] shows that it is always possible to *combine* algorithms for different possible situations into a single algorithm – even when it is not always algorithmically possible to decide which of the possible situations we currently encounter.

How to design efficient algorithms for solving the problems. There are several ways to design more efficient algorithms.

– First, it is often beneficial to *reformulate* the original problem.
  • In [13], it is shown that often real-world problems become easier to solve if we reformulate them in terms of constraints – e.g., in terms of constraint optimization – and then use constraint techniques to solve these problems.
  • Moreover, it turns out that sometimes, adding *additional* constraints [11] make problems easier to solve – constraints which, at first glance, would make the problem more complex to solve.
– Once the problem is formulated, we can try to come up with *more efficient algorithms* for solving the problem. This can be done both on a higher level – by coming up with a better numerical algorithm, or on lower level – by making elementary steps of the corresponding numerical algorithm more efficient. This volume contains examples of both approaches:
  • Innovative efficient algorithms for constraint optimization and equation solving are presented in [6, 19].

- Efficient algorithms for dealing with matrix and, more generally, tensor data are presented in [15].
- Often, while we do not have efficient general algorithms for solving a practical problem, human experts efficiently solve this problem. In such situations, it is important to learn how humans make decisions.
  - For multi-agent decision making in multi-criteria situations, such an analysis is presented in [5].
  - Similar studies of human decision making are also important in situations when we need to *influence* collective human decisions – e.g., evacuation in emergency situations [20].

How to take uncertainty into account. For a single variable, the simplest type of uncertainty is when we have bounds on this variables, i.e., when possible values of this variable form an *interval*. For interval uncertainty,

- a new more efficient method is described in [19]; this method is useful in solving systems of equations and in solving optimization problems under interval uncertainty;
- a new control techniques under interval uncertainty is described in [16].

In multi-dimensional case, in addition to intervals restricting the values of each variable, we may have additional constraints which limit the range of possible values of the corresponding tuples. Ellipsoids are often a computationally efficient tool for describing the resulting tools. The paper [23] provides a theoretical explanation for this empirical success. In [17], it is shown that sometimes half-ellipsoids provide an even more computationally efficient description of uncertainty.

Several papers take into account *probabilistic* uncertainty. The paper [18] uses interval techniques to simulate non-standard probability distribution useful in biological applications. The paper [9] shows that constraints techniques, when applied to statistical situations, explain well-known techniques of computational statistics such as Gibbs sampling. Finally, the paper [22] describes a new approach to solving problems with probabilistic uncertainty in which, in addition to continuous variables, more difficult-to-process discrete variables also need to be taken into account.

Constraint optimization problems under interval-valued *fuzzy* uncertainty are discussed in [4].

Resulting applications. Papers presented in this volume includes numerous applications. We want to emphasize three such applications:

- to meteorology and environmental science [7] (selecting the best location for a meteorological tower),
- to biology [18]: how to find the most probable evolution history of different species, and
- to engineering [16]: how to best control a magnetic levitation train.

Thanks. We are greatly thankful to National Science Foundation for supporting several CoProd workshops, to all the authors and referees, and to all the participants of the CoProd workshops. Our special thanks to Professor Janusz Kacprzyk, the editor of this book series, for his support and help. Thanks to all of you!

References

1. E. C. Balreira, O. Kosheleva, and V. Kreinovich, "Algorithmics of Checking Whether a Mapping Is Injective, Surjective, and/or Bijective", this volume.
2. M. Ceberio, O. Kosheleva, and V. Kreinovich, "Simplicity is worse than theft: a constraint-based explanation of a seemingly counter-intuitive Russian saying", this volume.
3. M. Ceberio and V. Kreinovich, "Continuous if-then statements are computable", this volume.
4. J. C. Figueroa-Garcia and G. Hernandez, "Linear programming with interval type-2 fuzzy constraints", this volume.
5. L. Garbayo, "Epistemic considerations on expert disagreement, normative justification and inconsistency regarding multi-criteria decision-making", this volume.
6. M. Hladik and J. Horacek, "Interval linear programming techniques in constraint programming and global optimization", this volume.
7. A. Jaimes, C. Tweedie, T. Magoc, V. Kreinovich, and M. Ceberio, "Selecting the best location for a meteorological tower: a case study of multi-objective constraint optimization", this volume.
8. A. Jalal-Kamali, M. Ceberio, and V. Kreinovich, "Constraint optimization: from efficient computation of what can be achieved to efficient computation of a way to achieve the corresponding optimum", this volume.
9. M. Koshelev, "Gibbs sampling as a natural statistical analog of constraints techniques: prediction in science under general probabilistic uncertainty", this volume.
10. O. Kosheleva, M. Ceberio, and V. Kreinovich, "Why tensors?", this volume.
11. O. Kosheleva, M. Ceberio, and V. Kreinovich, "Adding constraints – a (seemingly counterintuitive but) useful heuristic in solving difficult problems", this volume.
12. V. Kreinovich, "Under physics-motivated constraints, generally-non-algorithmic computational problems become algorithmically solvable", this volume.
13. V. Kreinovich, J. Ferret, and M. Ceberio, "Constraint-related reinterpretation of fundamental physical equations can serve as a built-in regularization", this volume.
14. T. Magoč and F. Modave, "Optimization of the Choquet Integral using Genetic Algorithm", this volume.
15. L. Mullin and J. Raynolds, "Scalable, portable, verifiable Kronecker products on multi-scale computers", this volume.
16. P. S. V. Nataraj and Mukesh D. Patil, "Reliable and Robust Automated Synthesis of QFT Controller for Nonlinear Magnetic Levitation System using Interval Constraint Satisfaction Techniques", this volume.
17. P. Portillo, M. Ceberio, and V. Kreinovich, "Towards an efficient bisection of ellipsoids", this volume.
18. R. Sainudiin, "An auto-validating rejection sampler for differentiable arithmetic expressions: posterior sampling of phylogenetic quartets", this volume.
19. S. Shary, "Graph subdivision methods in interval global optimization", this volume.

August 2013   Martine Ceberio
                                                                Vladik Kreinovich
                                                      University of Texas at El Paso

Table of Contents

Algorithmics of Checking whether a Mapping Is Injective, Surjective, and/or Bijective

E. Cabral Balreira[1], Olga Kosheleva[2], and Vladik Kreinovich[2]

[1] Department of Mathematics, Trinity University
San Antonio, TX 78212 USA
ebalreir@trinity.edu
[2] University of Texas at El Paso, El Paso, TX 79968, USA
{olgak,vladik}@utep.edu

Abstract. In many situations, we would like to check whether an algorithmically given mapping $f : A \to B$ is injective, surjective, and/or bijective. These properties have a practical meaning: injectivity means that the events of the action f can be, in principle, reversed, while surjectivity means that every state $b \in B$ can appear as a result of the corresponding action. In this paper, we discuss when algorithms are possible for checking these properties.

1 Formulation of the Problem

States of real-life systems change with time. In some cases, this change comes "by itself", from laws of physics: radioactive materials decays, planets go around each other, etc. In other cases, the change comes from our interference: e.g., a spaceship changes trajectory after we send a signal to an engine to perform a trajectory correction. In many situations, we have equations that describe this change, i.e., we know a function $f : A \to B$ that transform the original state $a \in A$ into a state $f(a) \in b$ at a future moment of time. In such situations, the following two natural problems arise.

The first natural question is: Are the changes reversible? For example, when we erase the value of the variable in a computer, by replacing it with 0s, the changes are not reversible: there is not trace of the original value left, and so reconstructing the original value is not possible. In such situations, two different original states $a \neq a'$ leads to the exact same new state $f(a) = f(a')$. If different states $a \neq a'$ always lead to different states $f(a) \neq f(a')$, then, in principle, we can reconstruct the original state a based on the new state $f(a)$. In mathematical terms, mapping $f : A \to B$ that map different elements into different ones are called *injective*, so the question is whether a given mapping is injective.

The second natural question is: Are all the states $b \in B$ possible as a result of this dynamics, i.e., is it true that every state $b \in B$ can be obtained as $f(a)$ for some $a \in A$. In mathematical terms, mappings that have this property are called *surjective*.

We may also want to check whether a mapping is both injective and surjective, i.e., in mathematical terms, whether it is a *bijection*.

M. Ceberio and V. Kreinovich (eds.), *Constraint Programming and Decision Making*,
Studies in Computational Intelligence 539,
DOI: 10.1007/978-3-319-04280-0_1, © Springer International Publishing Switzerland 2014

Thus, in practice, it is important to be able to check whether a given mapping is injective, surjective, or bijective; see, e.g., [1–3]. In this paper, we analyze this problem from an algorithmic viewpoint.

2   Case of Polynomial and, More Generally, Semi-algebraic Mappings

Case Study. Let us first consider the case when the set A and B are *semi-algebraic* sets, i.e., when each of these sets is characterized by a finite collection of polynomial equalities and inequalities with rational coefficients. For example, the upper half of the unit circle centered at the point $(0,0)$ is a semi-algebraic set, since it can be described as the set of all the pairs (x_1, x_2) that satisfy two polynomial inequalities: $x_1^2 + x_2^2 \leq 1$ and $x_2 \geq 0$.

We also assume that the mapping $f : A \to B$ is semi-algebraic – in the sense that the graph $\{(a, f(a)) : a \in A\}$ of this function is a semi-algebraic set. For example, every polynomial mapping is, by definition, semi-algebraic. Polynomial mappings are very important, since every continuous function on bounded set can be, within any given accuracy, approximated by a polynomial. Since in practice, we only know the actual consequences of each action with some accuracy, this means that every action can be represented by a polynomial mapping.

First Result: Algorithms Are Possible. In the polynomial case and, more generally, in the semi-algebraic case, all three above questions are algorithmically decidable:

Proposition 1. *There exists an algorithm, that, given two semi-algebraic sets A and B and a semi-algebraic mapping $f : A \to B$, checks whether f is injective, surjective, and/or bijective.*

Proof. Under the conditions of the proposition, each of the relations $a \in A$, $b \in B$, and $f(a) = b$ can be described by a finite set of polynomial equalities and inequalities. A polynomial is, by definition, a composition of additions and multiplications. Thus, both the injectivity and surjectivity can be described in terms of the first order language with variables running over real numbers, and elementary formulas coming from addition, multiplication, and equality. Namely, injectivity can be described as

$$\forall a \, \forall a' \, \forall b \, ((a \in A \,\&\, a' \in A \,\&\, f(a) = b \,\&\, f(a') = b \,\&\, b \in B) \to a = a'),$$

and surjectivity can be described as $\forall b \, (b \in B \to \exists a \, (a \in A \,\&\, f(a) = b))$. For such formulas, there is an algorithm – originally proposed by Tarski and later modified by Seidenberg – that decides whether a given formula is true or not; see, e.g., [5, 9, 12]. Thus, our problems are indeed algorithmically decidable.

Remark 1. One of the main open problems in this area is Jacobian Conjecture, according to which every polynomial map $f : C^n \to C^n$ from n-dimensional complex space into itself for which the Jacobi determinant $\det\left(\dfrac{\partial f_i}{\partial x_j}\right)$ is equal to 1 is injective; see, e.g., [4]. This is an open problem, but for any given dimension n and for any given degree d of the polynomial, the validity of the corresponding case of this conjecture can be resolved by applying the Tarski-Seidenberg algorithm.

How Efficient Are the Corresponding Algorithms? The following results show that the existence of the above algorithms do not mean that these algorithms are necessary efficient, even for polynomial mappings.

Proposition 2. *The problem of checking whether a given polynomial mapping* $f : \mathbb{R}^n \to \mathbb{R}^n$ *is injective is, in general, NP-hard.*

Proof. By definition, a problem is NP-hard if every problem from the class NP can be reduced to it; see, e.g., [10]. Thus, to prove that this problem is NP-hard, let us reduce a known NP-hard problem to it. As such a known problem, we take a *subset problem*: given $n + 1$ positive integers $s_1, \ldots, s_n, S$, check whether there exist values $\varepsilon_1, \ldots, \varepsilon_n \in \{0, 1\}$ for which $\sum_{i=1}^{n} s_i \cdot \varepsilon_i = S$. For each instance of this problem, let us form the following polynomial mapping $f(x_1, \ldots, x_n, x_{n+1}) = (x_1, \ldots, x_n, P(x_1, \ldots, x_n) \cdot x_{n+1})$, where

$$P(x_1, \ldots, x_n) \overset{\text{def}}{=} \sum_{i=1}^{n} x_i^2 \cdot (1 - x_i)^2 + \left(\sum_{i=1}^{n} s_i \cdot x_i - S \right)^2 .$$

If the original instance of the subset sum problem has a solution $(x_1, \ldots, x_n)$, then for this solution, we have $P(x_1, \ldots, x_n) = 0$ and thus, vectors

$$(x_1, \ldots, x_n, 0) \neq (x_1, \ldots, x_n, 1)$$

are mapped into the same vector $(x_1, \ldots, x_n, 0)$; so, f is not injective.

Vice versa, if the original instance of the subset sum problem does not have a solution, then $P(x_1, \ldots, x_n)$ is always positive – otherwise, the tuple $(\varepsilon_1, \ldots, \varepsilon_n) = (x_1, \ldots, x_n)$ would be a solution to this original instance. Thus, once we know

$$y = (y_1, \ldots, y_n, y_{n+1}) = f(x_1, \ldots, x_n, x_{n+1}),$$

we can recover all the inputs $x_1, \ldots, x_n, x_{n+1}$ as follows:

- $x_i = y_i$ for $i \leq n$ and
- $x_{n+1} = \dfrac{y_{n+1}}{P(x_1, \ldots, x_n)}.$

So, the above mapping f is injective if and only if the original instance of the subset problem has a solution. The reduction is proven, so the problem of checking injectivity is indeed NP-hard.

Proposition 3. *The problem of checking whether a given polynomial mapping* $f : \mathbb{R}^n \to \mathbb{R}^n$ *is surjective is, in general, NP-hard.*

Proof. This is proven by the same reduction as in the previous proof: when $P(x_1, \ldots, x_n) = 0$ for some $x_1, \ldots, x_n$, then the element $(x_1, \ldots, x_n, 1)$ is not in the range of the mapping; on the other hand, when P is always positive, the mapping is surjective.

Proposition 4. *The problem of checking whether a given polynomial mapping $f : \mathbb{R}^n \to \mathbb{R}^n$ is bijective is, in general, NP-hard.*

Proof. This is proven by the same reduction as in the previous two proofs.

Proposition 5. *The problem of checking whether a given surjective polynomial mapping $f : \mathbb{R}^n \to \mathbb{R}^n$ is also injective is, in general, NP-hard.*

Proof. Similarly to the proof of Proposition 2, for each instance $(s_1, \ldots, s_n, S)$ of the subset sum problem, we form the corresponding polynomial $P(x_1, \ldots, x_n)$. This polynomial is always non-negative. Let us prove that $P(x_1, \ldots, P_n)$ can attain values smaller than $\delta^2 \cdot (1 - \delta)^2$, where $\delta \overset{\text{def}}{=} \dfrac{0.1}{\displaystyle\sum_{i=1}^{n} |s_i|}$ if and only if the original instance of the subset sum problem has a solution.

Indeed, if the original instance of the subset sum problem has a solution, then, as we have shown in the proof of Proposition 2, the polynomial $P(x_1, \ldots, x_n)$ attains the 0 value for some inputs $x_1, \ldots, x_n$), and 0 is clearly smaller than $\delta^2 \cdot (1 - \delta)^2$. Vice versa, let us assume that for some inputs $x_1, \ldots, x_n$, we get $P(x_1, \ldots, x_n) < \delta^2 \cdot (1 - \delta)^2$. Since $P(x_1, \ldots, x_n)$ is the sum of non-negative terms, each of these terms must be smaller than $\delta^2 \cdot (1 - \delta)^2$. Each of these terms is a square q^2 of some expression q, so for each such expression q, we get $|q| < \delta \cdot (1 - \delta)$. In particular, for each i, we have $|x_i \cdot (1 - x_i)| < \delta \cdot (1 - \delta)$ and we also have $\left| \sum_{i=1}^{n} s_i \cdot x_i - S \right| < \delta \cdot (1 - \delta)$.

The inequality $-\delta \cdot (1-\delta) < x_i \cdot (1-x_i) < \delta \cdot (1-\delta)$ implies that either $|x_i| < \delta$ or $|x_i - 1| < \delta$, i.e., that there exists a value $\varepsilon_i \in \{0, 1\}$ for which $|x_i - \varepsilon_i| < \delta$. Thus, $|s_i \cdot x_i - s_i \cdot \varepsilon_i| < |s_i| \cdot \delta$, and $\left| \sum_{i=1}^{n} s_i \cdot x_i - \sum_{i=1}^{n} s_i \cdot \varepsilon_i \right| < \sum_{i=1}^{n} \cdot |s_i| \delta = 0.1$. From this inequality and the above inequality $\left| \sum_{i=1}^{n} s_i \cdot x_i - S \right| < \delta \cdot (1-\delta)$, we conclude that $\left| \sum_{i=1}^{n} s_i \cdot \varepsilon_i - S \right| < 0.1 + \delta \cdot (1 - \delta)$. Here, $\sum_{i=1}^{n} |s_i| \geq 1$, hence $0 \leq \delta \leq 0.1$ and $\delta \cdot (1-\delta) \leq \delta \leq 0.1$. Thus, $\left| \sum_{i=1}^{n} s_i \cdot \varepsilon_i - S \right| < 0.2$. Both the sum $\sum_{i=1}^{n} s_i \cdot \varepsilon_i$ and the value S are integers, so their difference is also an integer, and the only way for the absolute value of this difference to be smaller than 0.2 is when this difference is equal to 0, i.e., when $\sum_{i=1}^{n} s_i \cdot \varepsilon_i = S$. Thus, if $P(x_1, \ldots, x_n) < \delta^2 \cdot (1 - \delta)^2$ for some inputs x_i, then the original instance of the subset sum problem indeed has a solution.

For each instance of the subset sum problem, let us now use the corresponding polynomial $P(x_1, \ldots, x_n)$ to form the following polynomial mapping

$$f(x_1, \ldots, x_n, x_{n+1}) = (x_1, \ldots, x_n, x_{n+1}^3 + (P(x_1, \ldots, x_n) - \delta^2 \cdot (1 - \delta)^2) \cdot x_{n+1}).$$

This mapping maps each set of tuples with given $x_1, \ldots, x_n$ into the same set, so to check whether this mapping is surjective or injective, it is sufficient to check whether each corresponding 1-D mapping

$$g(x_{n+1}) \stackrel{\text{def}}{=} x_{n+1}^3 + (P(x_1, \ldots, x_n) - \delta^2 \cdot (1 - \delta)^2) \cdot x_{n+1}$$

is, correspondingly, surjective or injective.

A function $g(z) = z^3 + a \cdot z$ is always surjective: its values range from $-\infty$ for $z \to -\infty$ to $+\infty$ for $z \to +\infty$. When $a \geq 0$, this function is always increasing (hence injective), since its derivative $3z^2 + a$ is always non-negative. When $a < 0$, its derivative at 0 is negative, but this derivative is positive when $z \to \pm\infty$, so the function $g(z)$ is not monotonic and thus, not injective. So, the above mapping $f(x_1, \ldots, x_n, x_{n+1})$ is injective if and only if the coefficient $a = P(x_1, \ldots, x_n) - \delta^2 \cdot (1 - \delta)^2$ is non-negative for all $x_1, \ldots, x_n$, i.e., if and only if $P(x_1, \ldots, x_n) \geq \delta^2 \cdot (1 - \delta)^2$ for all $x_1, \ldots, x_n$. We have already shown that checking whether this inequality is always true is equivalent to checking whether the original instance of the subset sum problem has a solution. The reduction is proven, so the problem of checking injectivity of surjective mappings is indeed NP-hard.

Remark 2. It would be interesting to find out whether it is NP-hard to check whether a given injective polynomial mapping is bijective.

Polynomial Mapping with Computable Coefficients. For such mappings, the corresponding questions become algorithmically undecidable. A real number x is called *computable* if there exists an algorithm that, given a natural number n, returns a rational number r_n which is 2^{-m}-close to x. Equivalently, instead of specifying the sequence 2^{-n}, we can require the existence of an algorithm that, given a rational number $\varepsilon > 0$, produces a rational number which is ε-close to x; see, e.g., [6, 8, 11, 13].

Proposition 6. *No algorithm is possible that, given a polynomial mapping* $f : \mathbb{R}^n \to \mathbb{R}^n$ *with computable coefficients, decides whether this mapping is injective.*

Proof. The proof is based on the known fact that no algorithm is possible that, given a computable real number a, decides whether this number is equal to 0 or not. We can thus take $n = 1$ and $f(x) = a \cdot x$. This mapping is injective if and only if $a \neq 0$. Since we cannot algorithmically decide whether $a \neq 0$, we thus cannot algorithmically check whether a given mapping is injective.

Proposition 7. *No algorithm is possible that, given a polynomial mapping* $f : \mathbb{R}^n \to \mathbb{R}^n$ *with computable coefficients, decides whether this mapping is surjective.*

Proposition 8. *No algorithm is possible that, given a polynomial mapping* $f : \mathbb{R}^n \to \mathbb{R}^n$ *with computable coefficients, decides whether this mapping is bijective.*

Proof. These two results are proven by the same reduction as the previous proposition.

Proposition 9. *No algorithm is possible that, given an injective polynomial mapping* $f : [0,1] \to [0,1]$ *with computable coefficients, decides whether this mapping is also surjective.*

Proof. Indeed, for all $a \in [0, 0.5]$, the mapping $f(x) = (1 - a^2) \cdot x$ is injective, but it is surjective only for $a = 0$.

Proposition 10. *No algorithm is possible that, given an surjective polynomial mapping* $f : \mathbb{R}^n \to \mathbb{R}^n$ *with computable coefficients, decides whether this mapping is also injective.*

Proof. Indeed, for $n = 1$, the mapping $f(x) = -a^2 \cdot x^2 + x^3$ is always surjective, but it is injective only when $a^2 = 0$, i.e., when $a = 0$.

3   General Case

Analytical Expressions. If instead of allowing computable numbers, we allow general analytical expressions, i.e., expression in terms of elementary constants such as π and elementary functions such as sin, the above problems remain algorithmically undecidable. Indeed, according to Matiyasevich's solution of the tenth Hilbert problem, it is not algorithmically possible to check whether a given polynomial equality $F(x_1, \ldots, x_n) = 0$ has an integer solution. Thus, we can form a function as in the proof of Propositions 2, 3, and 4, with

$$P(x_1, \ldots, x_n) = \sum_{i=1}^{n} \sin^2(\pi \cdot x_i) + F^2(x_1, \ldots, x_n).$$

Here, $P = 0$ if and only if the equation $F = 0$ has an integer solution.

General Computable Case. For a computable mapping f between computable compact sets A and B [6, 13], we can efficiently check *approximate* injectivity and surjectivity. For example, instead of checking whether $f(a) = f(a')$ implies $a = a'$, we can check, for given $\varepsilon > 0$ and $\delta > 0$, whether $d(f(a), f(a')) \leq \delta$ implies $d(a, a') \leq \varepsilon$, i.e., whether

$$m \stackrel{\text{def}}{=} \max\{d(a, a') : d(f(a), f(a')) \leq \delta\} \leq \varepsilon.$$

It is known that between every two values $0 \leq \underline{\delta} < \overline{\delta}$, there exists a δ for which the set $\{d(f(a), f(a')) \leq \delta\}$ is a computable compact [6] and thus, for which m is computable. Thus, if we have two computable numbers $0 \leq \underline{\varepsilon} < \overline{\varepsilon}$, we can check whether $m \geq \overline{\varepsilon}$ or $m \not\geq \underline{\varepsilon}$. So, within each two intervals $(\underline{\delta}, \overline{\delta})$ and $(\underline{\varepsilon}, \overline{\varepsilon})$, we can algorithmically find values δ and ε for which the question of (δ, ε)-injectivity is algorithmically decidable.

For surjectivity, a natural idea is to check whether every $b \in B$ is ε-close to some $f(a)$, i.e., where $s \overset{\text{def}}{=} \max_{b \in B} \min_{a \in A} d(b, f(a)) \leq \varepsilon$. For computable mappings, s is computable, thus, with each interval $(\underline{\varepsilon}, \overline{\varepsilon})$, we can algorithmically find a value ε for which the question of ε-surjectivity is algorithmically decidable.

Acknowledgments. This work was supported in part by the National Science Foundation grants HRD-0734825 and DUE-0926721, and by Grant 1 T36 GM078000-01 from the National Institutes of Health.

References

1. Balreira, E.C.: Foliations and global inversion. Commentarii Mathematici Helvetici 85(1), 73–93 (2010)
2. Balreira, E.C.: Incompressibility and global inversion. Topological Methods in Nonlinear Analysis 35(1), 69–76 (2010)
3. Balreira, E.C., Radulescu, M., Radulescu, S.: A generalization of the Fujisawa-Kuh Global Inversion Theorem. Journal of Mathematical Analysis and Applications 385(2), 559–564 (2011)
4. Bass, H., Connell, E.H., Wright, D.: The Jacobian Conjecture: reduction of degree and formal expansion of the inverse. Bull. Amer. Math. Soc. 7(2), 287–330 (1982)
5. Basu, S., Pollack, R., Roy, M.-F.: Algorithms in Real Algebraic Geometry. Springer, Berlin (2006)
6. Bishop, E., Bridges, D.S.: Constructive Analysis. Springer, N.Y (1985)
7. Keller, O.-H.: Ganze Cremona-Transformationen. Monatshefte für Mathematik und Physik 47(1), 299–306 (1939)
8. Kreinovich, V., Lakeyev, A., Rohn, J., Kahl, P.: Computational complexity and feasibility of data processing and interval computations. Kluwer, Dordrecht (1998)
9. Mishra, B.: Computational real algebraic geometry. In: Handbook on Discreet and Computational Geometry. CRC Press, Boca Raton (1997)
10. Papadimitriou, C.H.: Computational Complexity. Addison Wesley, San Diego (1994)
11. Pour-El, M.B., Richards, J.I.: Computability in Analysis and Physics. Springer, Berlin (1989)
12. Tarski, A.: A Decision Method for Elementary Algebra and Geometry, 2nd edn. Berkeley, Los Angeles (1951)
13. Weihrauch, K.: Computable Analysis. Springer, Berlin (2000)

Simplicity Is Worse Than Theft:
A Constraint-Based Explanation of a Seemingly Counter-Intuitive Russian Saying

Martine Ceberio, Olga Kosheleva, and Vladik Kreinovich

University of Texas at El Paso, El Paso, TX 79968, USA
{mceberio,olgak,vladik}@utep.edu

Abstract. In many practical situations, simplified models, models that enable us to gauge the quality of different decisions reasonably well, lead to far-from-optimal situations when used in searching for an optimal decision. There is even an appropriate Russian saying: simplicity is worse than theft. In this paper, we provide a mathematical explanation of this phenomenon.

In Science, Simplicity Is Good. The world around us is very complex. One of the main objectives of science is to simplify it – and since has indeed greatly succeeded in doing it. For example, when Isaac Newton discovered his dynamical equations, it allowed him to explain the complex pictures of celestial bodies motion in terms of simple laws, laws that enable us to predict their positions for hundreds of years ahead.

From this viewpoint, a simplicity of a description is desirable. Yes, to achieve this simplicity, we sometimes ignore minor factors – but without such simplifying assumptions, it is difficult to make predictions, and predictions made based on these simplifying assumptions is usually reasonably good. For example, in his celestial mechanics studies, Newton ignored the fact that the planets and the Sun have finite size, and treated them as points with mass. To some extent, this simplifying assumption was justified: the gravitational field of a rotationally symmetric body is indeed the same as the field generated by the point with the same mass. However, real bodies are not perfectly symmetric, and there is a small discrepancy between the actual field and Newton's approximate values.

In Real Life, Simplified Models – That Seem to Be Working Well for Prediction – Are Sometimes Disastrous When We Move to Decision Making. One of the main purposes of science is to explain the world, to be able to predict what is happening in the world. Once this understanding is reached, once we have acquired the knowledge about the world, we use this knowledge to come up with actions that would make the world a better place.

For example, once the scientists get a better understanding of how cracks propagate through materials, it helps engineers design more stable constructions. Once the scientists learn about the life cycle of viruses, it helps develop medicines that prevent and cure the diseases caused by these viruses.

M. Ceberio and V. Kreinovich (eds.), *Constraint Programming and Decision Making*,
Studies in Computational Intelligence 539,
DOI: 10.1007/978-3-319-04280-0_2, © Springer International Publishing Switzerland 2014

What happens sometimes is that the simplified models, models which have led to very accurate predictions, are not as efficient when we use them in decision making. Numerous examples of such an inefficiency can be found in the Soviet experiment with the global planning of economy; see, e.g., [11] and references therein. In spite of using latest techniques of mathematical economics, including the ideas of the Nobelist Wassily Leontieff [5] who started his research as a leading USSR economist, the results were sometimes disastrous.

For example, during the Soviet times, buckwheat – which many Russian like to eat – was often difficult to buy. A convincing after-the-fact explanation is based on the fact that if we describe the economy in too many details, the corresponding optimization problem becomes too complex to solve. To make it solvable, the problem has been simplified, usually by grouping several similar quantities together. For example, many different types of grains were grouped together into a single grain rubric. The corresponding part of the optimization task became as follows: optimize the overall grain production under the given costs. The problem is that for the same expenses, we can get slightly more wheat than buckwheat. As a result, when we optimize the overall grain production, buckwheat is replaced by wheat – and thus, the buckwheat production shrank.

A similar example related to transportation is described in [10]. One of the main symptoms of an inefficient use of trucks is that sometimes, instead of delivering goods, trucks spend too much time waiting to be loaded, or getting stuck in traffic. Another symptom is when a truck is under-loaded, so that a small load is inefficiently transported by an unnecessarily large truck. In view of these symptoms, a natural way to gauge the efficiency of a transportation company is to measures the amount of tonne-kilometers that it produced during a given time period. If this amount is close to the product of the overall truck capacity and the distance which can be covered during this time period, the company is more efficient; if this amount is much smaller, there is probably room for improvement. In view of this criterion, when the first efficient reasonably large-scale optimization algorithms appeared in the last 1960s, scientists decided to use these algorithms to find the optimal transportation schedule in and around Moscow – by optimizing the total number of tonne-kilometers. The program did find a mathematically optimal solution, but this solution was absurd: load full weight on all the trucks in the morning and let them go round and round the Circular Highway around Moscow :-(

In all these anecdotal examples, a simplified model – which works reasonably well in estimating the relative quality of the existing situations – leads to absurd solutions when used for optimization. Such situations were so frequent that there is a colorful Russian saying appropriate for this phenomenon: simplicity is worse than theft.

Question. There is an anecdotal evidence of situations in which the use of simplified models in optimization leads to absurd solutions. How frequent are such situations? Are they typical or rare?

To answer this question, let us analyze this question from the mathematical viewpoint.

Reformulating the Question in Precise Terms. In a general decision making problem, we have a finite amount of resources, and we need to distribute them between n possible tasks, so as to maximize the resulting outcomes. For example, a farmer has a finite amount of money, and we need to allocate them to different possible crops so as to maximize the income. Police has a finite amount of officers, and we need to allocate them to different potential trouble spots so as to minimize the resulting amount of crime, etc. In some practical problems, we have limitations on several different types of resources, but for simplicity, we will assume that all resources are of one type.

Let x_0 be the total amount of resources, let n be the total number of possible tasks, and let $x_1, \ldots, x_n$ be the amounts allocated to different tasks; then, $x_i \geq 0$ and $x_1 + \ldots + x_n = x_0$. Let $f(x_1, \ldots, x_n)$ be the outcome corresponding to the allocation $x = (x_1, \ldots, x_n)$. In many practical problems, the amount of resources is reasonably small. In such cases, we can expand the dependence $f(x_1, \ldots, x_n)$ in the Taylor series and keep only linear terms in this expansion. In this case, the objective function takes a linear form $f(x_1, \ldots, x_n) = c_0 + \sum_{i=1}^{n} c_i \cdot x_i$. The question is then to find the values $x_1, \ldots, x_n \geq 0$ for which the sum $c_0 + \sum_{i=1}^{n} c_i \cdot x_i$ is the largest possible under the constraint that $\sum_{i=1}^{n} x_i = x_0$.

What Does Simplification Means in This Formulation. For this problem, simplification – in the sense of the above anecdotal examples – means that we replace individual variables by their sum. This can be always done if for two variables x_i and x_j, the coefficients c_i and c_j are equal. In this case, the sum of the corresponding terms in the objective function takes the form $c_i \cdot x_i + c_j \cdot x_j = c_i \cdot (x_i + x_j)$, so everything depends only on the sum $x_i + x_j$ (and does not depend on the individual values of these variables).

Since this replacement can be done exactly when the coefficients c_i and c_j are equal, it makes sense to perform a similar replacement when the coefficients c_i and c_j are close to each other. In this case, we replace both coefficients c_i and c_j, e.g., by their average. Similarly, if we have several variables with similar coefficients c_i, we replace all these coefficients by the average value.

Not all the variables have similar coefficients. Let us assume that for all other variables x_k, we have already selected some values, so only the variables with similar coefficients remain. In this case, the objective problem reduces to optimizing the sum $\sum c_i \cdot x_i$ over remaining variables, and the constraint take the form $\sum x_i = X_0$, where X_0 is equal to x_0 minus the sum of already allocated resources. If we now rename the remaining variables as $x_1, \ldots, x_m$, we arrive at the following situation:

– the original problem is to maximize the sum $f(x_1, \ldots, x_m) = \sum_{i=1}^{m} c_i \cdot x_i$ under the constraint $\sum_{i=1}^{m} x_i = X_0$;

– for simplicity, we replace this original problem by the following one: maximize the sum $s(x_1, \ldots, x_m) = \sum_{i=1}^{m} c \cdot x_i$ under the constraint $\sum_{i=1}^{m} x_i = X_0$.

The Simplified Description Provides, in General, a Reasonable Estimate for the Objective Function. Let us first show that the question is to *estimate* the value of the objective function corresponding to given allocations $x = (x_1, \ldots, x_m)$, then the estimation provided by the simplified expression is reasonably accurate.

Indeed, due to many different factors, the actual values c_i differ from the average c. There are so many different reasons for this deviation, that it makes sense to assume that the deviations $\Delta c_i \stackrel{\text{def}}{=} c_i - c$ are independent identically distributed random variables, with mean 0 and some standard deviation σ. In this case, the approximation error $a \stackrel{\text{def}}{=} f(x_1, \ldots, x_m) - s(x_1, \ldots, x_m)$ takes the form $a = \sum_{i=1}^{m} \Delta c_i \cdot x_i$. Since all Δc_i are independent, with mean 0 and standard deviation σ, their linear combination a has mean 0 and standard deviation $\sigma[a] = \sigma \cdot \sqrt{\sum_{i=1}^{m} x_i^2}$. In particular, when the resources are approximately equally distributed between different tasks, i.e., $x_i \approx \dfrac{X_0}{m}$, this standard deviation is equal to $\sigma[a] = X_0 \cdot \dfrac{\sigma}{\sqrt{m}}$. The actual value of the objective function is approximately equal to $s(x_1, \ldots, x_m) = c \cdot \sum_{i=1}^{m} x_i = c \cdot X_0$. Thus, the relative accuracy of approximating f by s can be described as the ratio $\dfrac{\sigma[a]}{s} = \dfrac{\sigma}{c \cdot \sqrt{m}}$. When m is large, this ratio is small, meaning that this simplification indeed leads to a very accurate estimation.

For Optimization, the Simplified Objective Function Can Lead to Drastic Non-optimality. From the mathematical viewpoint, the above optimization problem is easy to solve: to get the largest gain $\sum_{i=1}^{m} c_i \cdot x_i$, we should allocate all the resources X_0 to the task that leads to the largest amount of gain per unit resource, i.e., to the task with the largest possible value c_i. In this case, the resulting gain is equal to $X_0 \cdot \max_{i=1,\ldots,m} c_i$.

On the other hand, for the simplified objective function, its value is the same no matter how we distribute the resources, and is equal to $X_0 \cdot c$. In this simplified problem, it does not matter how we allocate the resources between the tasks, so we can as well allocate them equally. In this case, the resulting gain is indeed equal to $X_0 \cdot c$.

For random variables, the largest value $\max c_i$ is often much larger than the average c; moreover, the larger the sample size m, the more probable it is that we will observe values which are much larger than the average. This is especially true for power-law distributions which are frequent in economics and finance;

see, e.g., [1–4, 6–9, 12–14]. These distributions have heavy tails, with a high probability of c_i exceeding the mean. Thus, the simplified model can indeed lead to very non-optimal solutions.

Acknowledgments. This work was supported in part by the National Science Foundation grants HRD-0734825 and DUE-0926721, and by Grant 1 T36 GM078000-01 from the National Institutes of Health.

References

1. Chakrabarti, B.K., Chakraborti, A., Chatterjee, A.: Econophysics and Sociophysics: Trends and Perspectives. Wiley-VCH, Berlin (2006)
2. Chatterjee, A., Yarlagadda, S., Chakrabarti, B.K.: Econophysics of Wealth Distributions. Springer-Verlag Italia, Milan (2005)
3. Farmer, J.D., Lux, T.: Applications of statistical physics in economics and finance. A Special Issue of the Journal of Economic Dynamics and Control 32(1), 1–320 (2008)
4. Gabaix, X., Parameswaran, G., Vasiliki, P., Stanley, H.E.: Understanding the cubic and half-cubic laws of financial fluctuations. Physica A 324, 1–5 (2003)
5. Leontieff, W.: Input-Output Economics. Oxford University Press, New York (1986)
6. Mantegna, R.N., Stanley, H.E.: An Introduction to Econophysics: Correlations and Complexity in Finance. Cambridge University Press, Cambridge (1999)
7. McCauley, J.: Dynamics of Markets, Econophysics and Finance. Cambridge University Press, Cambridge (2004)
8. Rachev, S.T., Mittnik, S.: Stable Paretian Models in Finance. Wiley Publishers, New York (2000)
9. Roehner, B.: Patterns of Speculation - A Study in Observational Econophysics. Cambridge University Press, Cambridge (2002)
10. Romanovsky, J.V.: Lectures on Mathematical Economics. St. Petersburg University, Russia (1972)
11. Shmelev, N., Popov, V.: The Turning Point. Doubleday, New York (1989)
12. Stanley, H.E., Amaral, L.A.N., Gopikrishnan, P., Plerou, V.: Scale invariance and universality of economic fluctuations. Physica A 283, 31–41 (2000)
13. Stoyanov, S.V., Racheva-Iotova, B., Rachev, S.T., Fabozzi, F.J.: Stochastic models for risk estimation in volatile markets: a survey. Annals of Operations Research 176, 293–309 (2010)
14. Vasiliki, P., Stanley, H.E.: Stock return distributions: tests of scaling and universality from three distinct stock markets. Physical Review E: Statistical, Nonlinear, and Soft Matter Physics 77(3) (2008), Publ. 037101

Continuous If-Then Statements Are Computable

Martine Ceberio and Vladik Kreinovich

Department of Computer Science,
University of Texas at El Paso,
500 W. University,
El Paso, TX 79968, USA
{mceberio,vladik}@utep.edu

Abstract. In many practical situations, we must compute the value of an if-then expression $f(x)$ defined as "if $c(x) \geq 0$ then $f_+(x)$ else $f_-(x)$", where $f_+(x)$, $f_-(x)$, and $c(x)$ are computable functions. The value $f(x)$ cannot be computed directly, since in general, it is not possible to check whether a given real number $c(x)$ is non-negative or non-positive. Similarly, it is not possible to compute the value $f(x)$ if the if-then function is discontinuous, i.e., when $f_+(x_0) \neq f_-(x_0)$ for some x_0 for which $c(x_0) = 0$.

In this paper, we show that if the if-then expression is continuous, then we can effectively compute $f(x)$.

Practical Need for If-Then Statements. In many practical situations, we have different models for describing a phenomenon:

- a model $f_+(x)$ corresponding to the case when a certain constraint $c(x) \geq 0$ is satisfied, and
- a model $f_-(x)$ corresponding to the case when this constraint is not satisfied, i.e., when $c(x) < 0$ (usually, the second model is also applicable when $c(x) \leq 0$).

For example, in Newton's gravitation theory, when we are interested in the gravitation force generated by a celestial body – i.e., approximately, a sphere of a certain radius R – we end up with two different formulas:

- a formula $f_+(x)$ that describes the force outside the sphere, i.e., where

$$c(x) \stackrel{\text{def}}{=} \|\vec{r}\| - R \geq 0,$$

and
- a different formula $f_-(x)$ that describes the force inside the sphere, i.e., where

$$c(x) = \|\vec{r}\| - R \leq 0.$$

M. Ceberio and V. Kreinovich (eds.), *Constraint Programming and Decision Making*,   15
Studies in Computational Intelligence 539,
DOI: 10.1007/978-3-319-04280-0_3, © Springer International Publishing Switzerland 2014

Towards a Precise Formulation of the Computational Problem. In such situations, we have the following problem:

- we know how to compute the functions $f_+(x)$, $f_-(x)$, and $c(x)$;
- we want to be able to compute the corresponding "if-then" function

$$f(x) \stackrel{\text{def}}{=} \text{if } c(x) \geq 0 \text{ then } f_+(x) \text{ else } f_-(x).$$

In general, we say that a function $f(x)$ is *computable* if there is an algorithm that, given an input x and a rational number $\varepsilon > 0$, produces a rational number r for which $|f(x) - r| \leq \varepsilon$.

In the above formulation, we assume that the function $c(x)$ is computable for all possible values x from a given set X, and that:

- the function $f_+(x)$ is computable for all values $x \in X$ for which $c(x) \geq 0$; and
- the function $f_-(x)$ is computable for all values $x \in X$ for which $c(x) \leq 0$.

Why This Problem Is Non-trivial. The value $f(x)$ cannot be computed directly, since in general, it is not possible to check whether a given real number $c(x)$ is non-negative or non-positive; see, e.g., [2, 3].

Discontinuous If-Then Statements Are Not Computable. It is known that every computable function is everywhere continuous; see, e.g., [3].

Thus, when the if-then function $f(x)$ is not continuous, i.e., when $f_+(x_0) \neq f_-(x_0)$ for some x_0 for which $c(x_0) = 0$, then the function $f(x)$ is not computable.

Our Main Result. In this paper, we show that in all other cases, i.e., when the if-then function $f(x)$ is continuous, it is computable.

Algorithm: Main Idea. The main idea behind our algorithm is that in reality, we have one of the three possible cases:

- case of $c(x) > 0$, when $f(x) = f_+(x)$;
- case of $c(x) < 0$, when $f(x) = f_-(x)$; and
- case of $c(x) = 0$, when $f(x) = f_+(x) = f_-(x)$.

Let us analyze these three cases one by one.

In the first case, let us compute $c(x)$ with higher and higher accuracy $\varepsilon = 2^{-k}$, $k = 1, 2, \ldots$ As soon as we reach the accuracy $2^{-k} < \dfrac{c(x)}{2}$, for which $c(x) > 2 \cdot 2^{-k}$, we get an approximation r_k for which $|c(x) - r_k| \leq 2^{-k}$, i.e., for which

$$r_k > c(x) - 2^{-k} \geq 2 \cdot 2^{-k} - 2^{-k} = 2^{-k}$$

and thus, $r_k > 2^{-k}$. Since we know that $c(x) \geq r_k - 2^{-k}$, we thus conclude that $c(x) > 0$.

Similarly, in the second case, if we compute $c(x)$ with higher and higher accuracy 2^{-k}, we will reach an accuracy $2^{-k} < \dfrac{|c(x)|}{2}$, for which the corresponding approximate value r_k satisfy the inequality $r_k < -2^{-k}$ and thus, we can conclude that $c(x) < 0$.

In the third case, since $f_+(x) = f_-(x)$, if we compute $f_+(x)$ and $f_-(x)$ with accuracy $\varepsilon > 0$, then the resulting approximate values r_+ and r_- satisfy the inequalities $|f(x) - r_+| = |f_+(x) - r_+| \le \varepsilon$ and $|f(x) - r_-| = |f_-(x) - r_-| \le \varepsilon$ and therefore,

$$|r_+ - r_-| \le |r_+ - f(x)| + |f(x) - r_-| \le \varepsilon + \varepsilon = 2\varepsilon.$$

Vice versa, if the inequality $|r_+ - r_-| \le 2\varepsilon$ is satisfied (even if we know nothing about $c(x)$), then in reality, the value $f(x)$ coincides wither with $f_+(x)$ or with $f_-(x)$.

In the first subcase, when $f(x) = f_+(x)$, we have

$$|f(x) - r_+| = |f_+(x) - r_+| \le \varepsilon$$

and

$$|f(x) - r_-| = |f_+(x) - r_-| \le |f_+(x) - r_+| + |r_+ - r_-| \le \varepsilon + 2\varepsilon = 3\varepsilon.$$

Thus, due to convexity of the absolute value, we have

$$|f(x) - \overline{r}| \le \frac{1}{2} \cdot (|f(x) - r_+| + |f(x) - r_-|) \le \frac{\varepsilon + 3\varepsilon}{2} = 2\varepsilon.$$

In the second subcase, when $f(x) = f_-(x)$, we have

$$|f(x) - r_-| = |f_-(x) - r_-| \le \varepsilon$$

and

$$|f(x) - r_+| = |f_-(x) - r_+| \le |f_-(x) - r_-| + |r_- - r_+| \le \varepsilon + 2\varepsilon = 3\varepsilon.$$

Thus, due to convexity of the absolute value, we have

$$|f(x) - \overline{r}| \le \frac{1}{2} \cdot (|f(x) - r_-| + |f(x) - r_+|) \le \frac{\varepsilon + 3\varepsilon}{2} = 2\varepsilon.$$

In both case, we have $|f(x) - \overline{r}| \le 2\varepsilon$. So, if we want to compute $f(x)$ with a given accuracy $\alpha > 0$, it is sufficient to find $\dfrac{\alpha}{2}$-approximations r_- and r_+ to $f_-(x)$ and $f_+(x)$ for which $|r_+ - r_-| \le \alpha$

Thus, we arrive at the following algorithm for computing the if-then function $f(x)$.

Resulting Algorithm. To compute $f(x)$ with a given accuracy α, we simultaneously run the following three processes:

- computing $c(x)$ with higher and higher accuracy $\varepsilon = 2^{-k}$, $k = 1, 2, \ldots$;
- computing $f_-(x)$ with accuracy $\dfrac{\alpha}{2}$; and
- computing $f_+(x)$ with accuracy $\dfrac{\alpha}{2}$.

Let us denote:

- the result of computing $c(x)$ with accuracy 2^{-k} by r,
- the result of the second process by r_-, and
- the result of the third process by r_+.

As we have mentioned in our analysis, eventually, one of the following three events will happen:

- either we find out that $r_k > 2^{-k}$; in this case we know that $(c(x) > 0$ and hence) the third process will finish, so we finish it and return r_+ as the desired α-approximation to $f(x)$;
- or we find out that $r_k < -2^{-k}$; in this case we know that $(c(x) < 0$ and hence) the second process will finish, so we finish it and return r_- as the desired α-approximation to $f(x)$;
- or we find out that $|r_+ - r_i| \leq \alpha$; in this case, we return $\bar{r} = \dfrac{r_- + r_+}{2}$ as the desired α-approximation to $f(x)$.

Historical Comment. Our proof is a simplified version of the proofs described, in a more general setting, in [3]; see also [1].

Acknowledgments. This work was supported in part by the National Science Foundation grants HRD-0734825 and DUE-0926721, and by Grant 1 T36 GM078000-01 from the National Institutes of Health.

References

1. Brattka, V., Gherardi, G.: Weihrauch degrees, omniscience principle, and weak computability. In: Bauer, A., Dillhage, R., Hertling, P., Ko, K.-I., Rettinger, R. (eds.) Proceedings of the Sixth International Conference on Computability and Complexity in Analysis CCA 2009, Ljubljana, Slovenia, August 18-22, pp. 81–92 (2009)
2. Kreinovich, V., Lakeyev, A., Rohn, J., Kahl, P.: Computational Complexity and Feasibility of Data Processing and Interval Computations. Kluwer, Dordrecht (1998)
3. Weihrauch, K.: Computable Analysis: An Introduction. Springer, New York (2000)

Linear Programming with Interval Type-2 Fuzzy Constraints

Juan C. Figueroa-García and Germán Hernández

Universidad Nacional de Colombia, Bogota, Colombia
jcfigueroag@udistrital.edu.co, gjhernandezp@gmail.com

Abstract. This chapter shows a method for solving Linear Programming (LP) problems that includes Interval Type-2 fuzzy constraints. A method is proposed for finding an optimal solution in these conditions, using convex optimization techniques. The entire method is presented and some interpretation issues are discussed. An introductory example is presented and solved using our proposal, and its results are explained and discussed.

1 Introduction

A special kind of Linear Programming (LP) models address fuzzy constraints, those models are known as Fuzzy Linear Programming (FLP) models. There are different ways for modeling fuzzy constraints, each one at a different complexity level. Roughly speaking, fuzzy constrained problems are interesting since fuzzy sets can deal with non-probabilistic uncertainty, which is a common practical issue.

Some FLP models have been defined by Ghodousiana and Khorram in [1], Sy-Min and Yan-Kuen in [2], Tanaka and Asai in [3], Tanaka, Okuda and Asai in [4], Inuiguchi in [5], [6] and [7] who proposed solutions for several linear fuzzy sets, all of them considering only Type-1 fuzzy sets. Recently, an intuitionistic fuzzy optimization approach have been presented by Angelov [8] and Dubey et al. [9], which is based on the idea of using two measures $\mu_A(x)$ and $\upsilon_A(x)$ to represent both membership and non-membership degrees of x regarding a concept A, constrained to $0 \leq \mu_A(x) + \upsilon_A(x) \leq 1$, which is similar to a special kind of Type-2 fuzzy set in the sense that the interval between $\mu_A(x)$ and $\upsilon_A(x)$ can be shown as its footprint of uncertainty.

This chapter presents an extension of the method proposed by Zimmermann [10] and [11], originally designed for Type-1 fuzzy constrained problems, to an Interval Type-2 fuzzy constrained LP with piecewise linear membership functions. The proposal is based on the use of Type-2 fuzzy numbers instead of intervals or intuitionistic fuzzy sets which are alternative representations of uncertainty.

The chapter is divided into seven sections. In Section 1 the introduction and motivation are presented. In Section 2 the classical LP model with fuzzy

M. Ceberio and V. Kreinovich (eds.), *Constraint Programming and Decision Making*,
Studies in Computational Intelligence 539,
DOI: 10.1007/978-3-319-04280-0_4, © Springer International Publishing Switzerland 2014

constraints is presented. In Section 3, some elements of linguistic uncertainty, in particular Type-2 fuzzy constrains are introduced. In Section 4, the formal definition of an Interval Type-2 FLP model is presented. In Section 5, the proposed optimization method is explained. Section 6 presents an illustrative application example, and finally in Section 7 some concluding remarks are presented.

1.1 Applicability of Type-2 Fuzzy Sets

In practical applications such as financial, supply chain, Markov chains, control, etc, the analyst needs to make a decision based on the decision variables x of the system, so optimization techniques are oriented to find their optimal values x^*, even when the problem is under uncertainty conditions.

In LP, all its parameters (costs, technological coefficients and constraints) are considered as constant, but in practice we have that they may contain uncertainty (randomness, fuzziness, etc). As more uncertainty sources are considered, more complex is the model and the method used to reach a solution. A common scenario appears when defining a constraint since it could contain noise, the method used for measuring may not be totally confident, or simply there is no available data to define it, so decision makers (mostly experts of the system) have to apply different strategies to handle uncertainty and find a solution of the problem.

When the constraints of the problem are defined by the opinion of multiple experts or they are based on non-probabilistic information, the problem is how to measure those opinions and linguistic judgements, and then try to solve the problem. Since 1960's, another kind of uncertainty called linguistic uncertainty has been defined. In this, the uncertainty about different perceptions of a concept, mostly given by multiple experts with equally valuable opinions affects the definition of the constraints of an LP problem. This kind of uncertainty can be addressed using Interval Type-2 Fuzzy Sets (IT2FS).

An IT2FS is a more complex measure, so it needs more complex representations than classical fuzzy sets. In this way, our proposal is based on reducing its complexity using a Type-reduction strategy that consists on finding a set embedded into a Type-2 fuzzy set, in order to apply convex optimization techniques.

2   Basic Definitions

The linear programming (LP) problem is the problem of finding, among all vectors x which satisfy a system of n inequalities $Ax \leqslant b$, the vector which attains the largest value of the given objective function $z = c'x$. Now, a *solution* of an LP is a vector x which simply satisfies all the constraints of the problem $Ax \leqslant b$, and its *optimal solution* is a vector x^* for which we have $\sup\{c'x \mid Ax \leqslant b, x \geqslant 0\}$, so $z(x^*) \geqslant z(x)$ for all $x \in \mathbb{B}$, where all its parameters are crisp numbers (a.k.a. constants or just numbers). For further information see Dantzig [12], and Bazaraa, Jarvis and Sherali [13].

A fuzzy set A is a generalization of a crisp number. It is defined over a universe of discourse X and is characterized by a *Membership Function* namely $\mu_A(x)$ that takes values in the interval $[0,1]$. A fuzzy set A may be represented as a set of ordered pairs consisting of a generic element x and its grade of membership function, $\mu_A(x)$, i.e.,

$$A = \{(x, \mu_A(x)) \,|\, x \in X\} \tag{1}$$

The classical FLP problem is solves n inequalities as well, but using fuzzy sets as boundaries, namely B with parameters $\check{b}$ and $\hat{b}$, which are typically defined by piecewise linear membership functions as shown in Figure 1. A fuzzy constraint is then a partial order $\lesssim$ for which we have $x \lesssim B$.

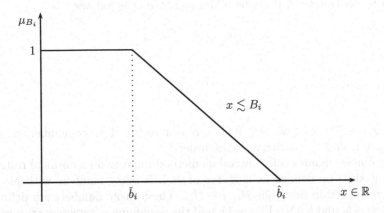

Fig. 1. Fuzzy set B_i

Now, the FLP problem is the problem of solving n inequalities of the system $Ax \lesssim B$, achieving the best value of given a goal $z = c'x$, and a solution of an FLP is a vector x which satisfies all the constraints of the problem $Ax \lesssim B$.

The concept of optimal solution of an FLP is different (but not far) from the optimality concept in LP. Fuzzy decision making basic principles were proposed by Bellman and Zadeh [14], where the main idea to obtain a maximum intersection among fuzzy constraints and fuzzy goals (Z_k), and then find a maximum fulfilment of all fuzzy parameters. Given this basis, the FLP problem becomes to a problem of finding a vector of solutions $x \in \mathbb{R}^m$ for a single goal (Z) such that:

$$\max_{x \in \mathbb{R}^m} {}^{\alpha}\left\{ \bigcap_{i=1}^{m} \{{}^{\alpha}B_i,\, b_i\} \bigcap Z \right\} \tag{2}$$

Note that this value α is an operation point for which the intersection among all constraints and a goal is maximum. This point α leads to a vector x^* for which we have to compute $\sup\{c'x \,|\, Ax \lesssim B, x \geqslant 0\}$, where $z(x^*) \geqslant z(x)\,\forall x \in \mathbb{B}$. In practice, this is a two goal problem where its first goal is to find a α which fulfills (2) and the second goal is to find x^* for which $z(x^*) \geqslant z(x)\,\forall x \in \mathbb{B}$.

Based on those principles, Zimmermann proposed a method for finding a solution, which is commonly called *Zimmermann soft constraints method* (See Zimmermann [10] and [11]). His proposal uses piecewise linear fuzzy numbers for defining B and an auxiliary variable α which follows the Bellman-Zadeh decision making principle, based on the fact that Z can be computed as a linear combination of the values of B, where $\check{z} = \sup\{c'x \mid Ax \leqslant \check{b}, x \geqslant 0\}$ and $\hat{z} = \sup\{c'x \mid Ax \leqslant \hat{b}, x \geqslant 0\}$ with linear piecewise membership function defined as the complement of μ_B. A brief explanation about this method is shown next.

2.1   The Zimmermann's Soft Constraints Model

The soft constraints FLP model addressed here is as follows:

$$\max_{x \in X} z = c'x + c_0$$

$$s.t.$$

$$Ax \lesssim B \tag{3}$$

$$x \geqslant 0$$

where $x, c \in \mathbb{R}^m$, $c_0 \in \mathbb{R}$, $A \in \mathbb{R}^{n \times m}$. B is a vector of fuzzy numbers as shown in Figure 1, and $\lesssim$ is a fuzzy partial order.[1]

The Zimmermann's soft constraints method imposes an additional restriction on B: it shall be defined as a vector of m L-R fuzzy numbers with piecewise linear membership functions $\tilde{B}_i$, $i \in \mathbb{N}_m$. These fuzzy numbers are defined by parameters $\check{b}_i$ and $\hat{b}_i$ (See Figure 1); and the remaining parameters are constants viewed as fuzzy singletons. Zimmermann proposed a method for solving this fuzzy constrained problems, described as follows:

1. Compute the inferior boundary of optimal solutions $\min\{z^*\} = \check{z}$ by using $\check{b}_i$ as a right hand side of the model.
2. Compute the superior boundary of optimal solutions $\max\{z^*\} = \hat{z}$ by using $\hat{b}_i$ as a right hand side of the model.
3. Define a fuzzy set $Z(x^*)$ with parameters $\check{z}$ and $\hat{z}$. This set represents the set of all feasible solutions regarding the objective. In other words, a thick solution of the fuzzy problem (See Kall and Mayer [15] and Mora [16]). Given the objective is to maximize, then its membership function is:

$$\mu_Z(x; \check{z}, \hat{z}) = \begin{cases} 1, & c'x \geqslant \hat{z} \\ \dfrac{c'x - \check{z}}{\hat{z} - \check{z}}, & \check{z} \leqslant c'x \leqslant \hat{z} \\ 0, & c'x \leqslant \check{z} \end{cases} \tag{4}$$

Its graphical representation is:

---

[1] Usually B is a linear fuzzy number, but there is the possibility to use nonlinear shapes.

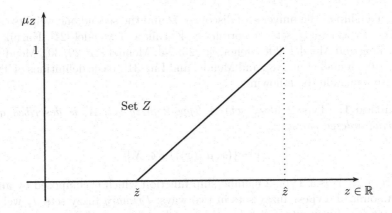

Fig. 2. Fuzzy set Z

4. Create an auxiliary variable α and solve the following model:

$$\max \{\alpha\}$$
$$s.t.$$
$$c'x + c_0 - \alpha(\hat{z} - \check{z}) = \check{z} \tag{5}$$
$$Ax + \alpha(\hat{b} - \check{b}) \leqslant \hat{b}$$
$$x \geqslant 0,\ \alpha \in [0, 1]$$

5. Return z^*, x^* and α^*.

This method uses α as a global satisfaction degree of all constraints regarding a fuzzy set of optimal solutions Z. In fact, α operates as a balance point between the use of the resources (denoted by the constraints of the problem) and the desired profits (denoted by z), since the use of more resources lead to more profits, at different uncertainty degrees. Then, the main idea of this method is to find an overall satisfaction degree of both goals (Profits vs. resource use) that maximizes the global satisfaction degree, i.e. minimizing the global uncertainty.

3   Interval Type-2 Fuzzy Constraints

As mentioned before, interval type-2 fuzzy sets allows to model lunguistic uncertainty, i.e. the uncertainty about different perceptions and concepts. Mendel [17–22] and Melgarejo [23, 24] provided formal definitions of IT2FS, and Figueroa [25–28] proposed an extension of the FLP to include constraints with linguistic uncertainty represented by IT2FS called Interval Type-2 Fuzzy Linear Programming (IT2FLP) which are shown next

3.1   Basics on Interval Type-2 Fuzzy Sets

A Type-2 fuzzy set is a collection of infinite Type-1 fuzzy sets into a single fuzzy set. It is defined by two membership functions: The first one defines the degree of membership of the universe of discourse Ω and the second one weights each of the first Type-1 fuzzy sets. According to Karnik and Mendel [21], Karnik et.al. [22], Liang and Mendel [20], Melgarejo [23, 24], Mendel [17, 29, 30], Mendel and John [19], Mendel et.al. [18], and Mendel and Liu [31], basic definitions of Type-2 fuzzy sets include the following:

Definition 1. (Type-2 fuzzy set) *A Type-2 fuzzy set, $\tilde{A}$, is described as the following ordered pairs:*

$$\tilde{A} = \{(x, \mu_{\tilde{A}}(x)) \,|\, x \in X\} \tag{6}$$

Here, $\mu_{\tilde{A}}(x))$ is a Type-2 membership function which is composed by an infinite amount of Type-1 fuzzy sets in two ways: *Primary* fuzzy sets J_x weighted by *Secondary* fuzzy sets $f_x(u)$. In other words

$$\tilde{A} = \{((x, u), J_x, f_x(u)) \,|\, x \in X; u \in [0, 1]\} \tag{7}$$

And finally we can get the following compact representation of $\tilde{A}$

$$\tilde{A} = \int_{x \in X} \int_{u \in J_x} f_x(u)/(x, u) = \int_{x \in X} \left[\int_{u \in J_x} f_x(u)/u \right] \Big/ x, \tag{8}$$

where x is the *primary variable*, J_x is the *primary membership function* associated to x, u is the *secondary variable*, and $\int_{u \in J_x} f_x(u)/u$ is the *secondary membership function*.

Uncertainty about $\tilde{A}$ is conveyed by the union of all of the primary memberships, called the *Footprint Of Uncertainty* of $\tilde{A}$ [FOU($\tilde{A}$)], i.e.

$$\text{FOU}(\tilde{A}) = \bigcup_{x \in X} J_x \tag{9}$$

Therefore, the FOU evolves all the embedded J_x weighted by the secondary membership function $f_x(u)/u$. These Type-2 fuzzy sets are known as *Generalized* Type-2 fuzzy sets, *(T2FS)*, since $f_x(u)/u$ is a Type-1 membership function. Now, an *Interval Type-2* fuzzy set, *(IT2FS)*, is a simplification of T2FS in the sense that the secondary membership function is assumed to be 1, as follows

Definition 2. (Interval Type-2 fuzzy set) *An Interval Type-2 fuzzy set, $\tilde{A}$, is described as:*

$$\tilde{A} = \int_{x \in X} \int_{u \in J_x} 1/(x, u) = \int_{x \in X} \left[\int_{u \in J_x} 1/u \right] \Big/ x, \tag{10}$$

While a T2FS uses any form of Type-1 membership functions, an IT2FS differs to a T2FS since it uses $f_x(u)/u = 1$ as a unique weight for each J_x, being an interval fuzzy set.

The FOU of an IT2FS is bounded by two membership functions: An *Upper* membership function *(UMF)* $\bar{\mu}_{\tilde{A}}(x)$ and a *Lower* membership function *(LMF)* $\underline{\mu}_{\tilde{A}}(x)$. A graphical representation is provided in Figure 2.

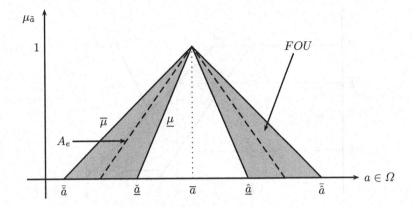

Fig. 3. Interval Type-2 Fuzzy set $\tilde{a}$

Here, $\tilde{a}$ is an Interval Type-2 fuzzy set defined over an universe of discourse $a \in \Omega$, its support $supp(\tilde{a})$ is enclosed into the interval $a \in [\bar{\bar{a}}, \hat{\bar{a}}]$. $\mu_{\tilde{a}}$ is a linear Type-2 fuzzy set with parameters $\bar{\bar{a}}, \hat{\bar{a}}, \check{\underline{a}}, \hat{\underline{a}}$ and $\bar{a}$, and A_e is a Type-1 fuzzy set embedded in the FOU.

There are many ways to define the *"knowledgeability"* of any expert, so an infinite number of A_e fuzzy sets are embedded in the FOU, each of these sets is a representation of either the the knowledge of an expert about the universe of discourse or his perception about it, which is an uncertainty source.

3.2 Uncertain Constraints

In this chapter, we refer to uncertain constraints when using IT2FS, so the presented approach solves interval type-2 fuzzy constraints (IT2FC). An IT2FC is a type-2 fuzzy partial order namely $\precsim$ for which we have $Ax \precsim \tilde{b}$. The membership function which represents $\tilde{b}_i$ is:

$$\tilde{b}_i = \int_{b_i \in \mathbb{R}} \left[\int_{u \in J_{b_i}} 1/u \right] \Big/ b_i, \ i \in \mathbb{N}_m, J_{b_i} \subseteq [0,1] \tag{11}$$

Note that $\tilde{b}$ is bounded by both *Lower* and *Upper* primary membership functions, namely $\underline{\mu}_{\tilde{b}}(x)$ with parameters $\underline{b}$ and $\hat{\underline{b}}$, and $\bar{\mu}_{\tilde{b}}(x)$ with parameters $\bar{\bar{b}}$ and $\hat{\bar{b}}$. Now, the FOU of each $\tilde{b}$ can be composed by two distances called $\triangle$ and ∇, where $\triangle$ is the distance between $\underline{b}$ and $\bar{\bar{b}}$, $\triangle = \bar{\bar{b}} - \underline{b}$ and ∇ is the distance between $\hat{\underline{b}}$ and $\hat{\bar{b}}$, $\nabla = \hat{\bar{b}} - \hat{\underline{b}}$. For further information about $\triangle$ and ∇ see Figueroa [26]. A graphical representation of $\tilde{b}_i$ is shown in Figure 4

In this Figure, $\tilde{b}$ is an IT2FS with linear membership functions $\underline{\mu}_{\tilde{b}}$ and $\bar{\mu}_{\tilde{b}}$. A particular value b projects an interval of infinite membership degrees $u \in J_b$, as follows

$$J_b \in \left[{}^{\alpha}\bar{b}, {}^{\alpha}\underline{b} \right] \ \forall \ b \in \mathbb{R} \tag{12}$$

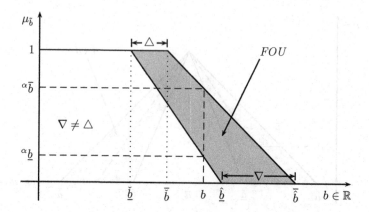

Fig. 4. IT2FS constraint with joint uncertain $\triangle$ & ∇

where J_b is the set of all possible membership degrees associated to $b \in \mathbb{R}$. Now, the FOU of $\tilde{b}$ can be composed by the union of all values of u, i.e.

Definition 3. (FOU of $\tilde{b}$) *Using* (12) *it is possible to compose the footprint of uncertainty of* $\tilde{b}, u \in J_b$ *as follows:*

$$\mathrm{FOU}(\tilde{b}) = \bigcup_{b \in \mathbb{R}} \left[{}^{\alpha}\overline{b}, \; {}^{\alpha}\underline{b} \right] \; \forall \; b \in \tilde{b}, \; u \in J_b, \; \alpha \in [0, 1] \tag{13}$$

Some interesting questions arise from the concept of an optimal solution in terms of the decision variables $x \in \mathbb{R}$ given uncertain constraints $\tilde{b}$. The natural way is by Type-reducing all IT2FS using centroid-based methods, and afterwards solve the resultant interval optimization problem. However, this is not recommendable because the centroid of an IT2FS constraint usually is outside its FOU. Another easy way is by using the *Center of FOU* which is simply to use the center of ∇ and $\triangle$ as extreme points of a fuzzy set embedded into the FOU of $\tilde{b}$, and then apply the Zimmermann's method. This method can be used in cases where the analyst has no complete knowledge about $\tilde{b}$.

In the following sections, some definitions about LP problems with IT2FS constraints are provided together with a method for finding optimal solutions in terms of $x \in \mathbb{R}$ regarding z and $\tilde{b}$.

4   The IT2FLP Model

Given the concept of an IT2FS constraint and the definition of an FLP, an uncertain constrained FLP model (IT2FLP) can be defined as follows:

$$\max_{x \in X} z = c'x + c_0$$

$$s.t.$$

$$Ax \precsim \tilde{b} \tag{14}$$

$$x \geqslant 0$$

where $x, c \in \mathbb{R}^m$, $c_0 \in \mathbb{R}$, $A \in \mathbb{R}^{n \times m}$. $\tilde{b}$ is an IT2FS vector defined by two primary membership functions $\underline{\mu}_b$ and $\bar{\mu}_b$. $\precsim$ is a Type-2 fuzzy partial order.

Two possible partial orders $\precsim$ and $\succsim$ can be used, depending on the nature of the problem. In our approach, only linear membership functions are used since the main goal is to use LP models, due to they are easy to be optimized using classical algorithms. The membership function of $\precsim$ is:

$$\underline{\mu}_{\tilde{b}}(x; \underline{\check{b}}, \underline{\hat{b}}) = \begin{cases} 1, & x \leqslant \underline{\check{b}} \\ \dfrac{\underline{\hat{b}} - x}{\underline{\hat{b}} - \underline{\check{b}}}, & \underline{\check{b}} \leqslant x \leqslant \underline{\hat{b}} \\ 0, & x \geqslant \underline{\hat{b}} \end{cases} \tag{15}$$

and its upper membership function is:

$$\bar{\mu}_{\tilde{b}}(x; \bar{\check{b}}, \bar{\hat{b}}) = \begin{cases} 1, & x \leqslant \bar{\check{b}} \\ \dfrac{\bar{\hat{b}} - x}{\bar{\hat{b}} - \bar{\check{b}}}, & \bar{\check{b}} \leqslant x \leqslant \bar{\hat{b}} \\ 0, & x \geqslant \bar{\hat{b}} \end{cases} \tag{16}$$

A first approach for solving IT2FS problems is by reducing its complexity into a simpler form, allowing to use well known algorithms. In this case, we propose a methodology where its first step is to compute a fuzzy set of optimal solutions namely $\tilde{z}$ and afterwards, a Type-reduction strategy to find an embedded Type-reduced fuzzy set Z, is applied. This allows us to find an optimal solution using the Zimmermann's method, so the above is currently the problem of finding a vector of solutions $x \in \mathbb{R}^m$ such that:

$$\max_{x \in \mathbb{R}^n} \alpha \left\{ \bigcap_{i=1}^{m} \{ {}^\alpha B_i, \, {}^\alpha \tilde{b}_i, \, b_i \} \bigcap \tilde{z} \right\} \tag{17}$$

where, ${}^\alpha B_i$ and ${}^\alpha \tilde{b}_i$ are α-cuts made over all fuzzy constraints.

Given $\mu_{\tilde{z}}$, the problem becomes in how to find the maximal intersection point between $\tilde{z}$ and $\tilde{b}$, for which α is defined as auxiliary variable. In practice, the problem is solved by x^*, so α allows us to find x^*, according to (17). The proposed methodology is presented in Figure 5.

4.1 Convexity of an IT2FLP

Another important condition to be satisfied by any LP model is *convexity*. In an LP problem, convexity is a concept which means that the halfspace generated by all $A(x_{ij}) \leqslant b$ should be continuous. This means that every set b should not be empty (non-null).

In an FLP, two convexity conditions should be guaranteed: a first one regarding $b \in supp(\tilde{b})$ which is supposed to be a convex space, and a second one regarding $\tilde{b}$. This leads us to the following proposition:

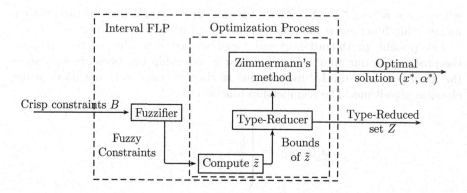

Fig. 5. IT2FLP proposed methodology

Proposition 1. (Convexity of an IT2FLP) *An IT2FLP is said to be convex iff*

$$A(x_{ij}) \precsim \tilde{b} \quad \forall \; i \in \mathbb{N}_m \tag{18}$$

is a non-null halfspace, and $\tilde{b}$ is composed by convex $\overline{\mu}_{\tilde{b}}$ and $\underline{\mu}_{\tilde{b}}$ membership functions.

Figueroa and Hernández [28] computed the set of all possible optimal solutions of an IT2FLP, as function of $\tilde{b}$. Based on Kreinovich et.al. [32], global optimization is only possible for convex objective functions, so the Proposition 1 states that the space of b should be defined by both a convex universe of discourse and membership functions ($\mu_{\underline{b}}$, $\mu_{\overline{b}}$). As $\tilde{z}$ is a function of $\tilde{b}$, $\tilde{b} \rightarrow \tilde{z}$, then $\tilde{b}$ need to be defined by a convex membership function to ensure that $\tilde{z}$ be also convex.

Remark 1 (Feasibility condition of the IT2FLP). The crisp boundaries of each set $\tilde{b}_i$ generates a halfspace $h(\cdot)$ defined as follows:

$$h(\cdot) \leqslant \tilde{\overline{b}}_i \quad \forall \; i \in \mathbb{N}_m \tag{19}$$

This means that $h(\cdot)$ is the halfspace generated by the set of all the values of x contained into the support of $\tilde{b}$, $x \in supp(\tilde{b})$ (See Niewiadomski [33, 34]). In this way, the IT2FLP model is feasible only if the polyhedron (or polytope) generated by $h(\cdot)$ is a non-trivial set, that is:

$$\mathcal{P} = \{x \mid h(\cdot) \leqslant \tilde{\overline{b}}_i, \} \tag{20}$$

where $\mathcal{P}$ is a non-trivial set of solutions (polyhedron or polytope) of a crisp LP model. Here, $\mathcal{P}$ is a convex set of solutions of all the inequalities of IT2FLP.

Therefore, the problem is feasible only if the broadest value contained into $\tilde{b}$ is feasible, i.e the one provided by $\bar{\tilde{b}}$. It is clear that if there exists a solution at this point, then all values of $b \leqslant \bar{\tilde{b}}$ are feasible as well, since they are contained into the convex hull defined by $\tilde{b}$ (See Wolsey [35], and Papadimitriou and Steiglitz [36]).

5   Solution Procedure of an IT2FLP

Figueroa [25–27] proposed a method that uses $\triangle$, ∇ as auxiliary variables with weights $c^{\triangle}$ and c^{∇} respectively, in order to find an optimal fuzzy set embedded into the FOU of the problem and then solve it by using the Zimmermann's method. Its description is presented next.

1. Calculate an optimal inferior boundary called Z $minimum$ ($\check{z}$) by using $\underline{\check{b}} + \triangle$ as a frontier of the model, where $\triangle$ is an auxiliary set of variables weighted by $c^{\triangle}$ which represents the lower uncertainty interval subject to the following statement:
$$\triangle \leqslant \bar{\tilde{b}} - \underline{\check{b}} \tag{21}$$

   To do so, $\triangle^*$ are obtained solving the following LP problem
$$\max_{x,\triangle} z = c'x + c_0 - c^{\triangle}{}'\triangle$$
$$s.t.$$
$$Ax - \triangle \leqslant \underline{\check{b}} \tag{22}$$
$$\triangle \leqslant \bar{\tilde{b}} - \underline{\check{b}}$$
$$x \geqslant 0$$

2. Calculate an optimal superior boundary called Z $maximum$ ($\hat{z}$) by using $\bar{\tilde{b}} + \nabla$ as a frontier of the model, where ∇ is an auxiliary set of variables weighted by c^{∇} which represents the upper uncertainty interval subject to the following statement:
$$\nabla \leqslant \bar{\tilde{b}} - \hat{\underline{b}} \tag{23}$$

   To do so, ∇^* are obtained solving the following LP problem
$$\max_{x,\nabla} z = c'x + c_0 - c^{\nabla}{}'\nabla$$
$$s.t.$$
$$Ax - \nabla \leqslant \hat{\underline{b}} \tag{24}$$
$$\nabla \leqslant \bar{\tilde{b}} - \hat{\underline{b}}$$
$$x \geqslant 0$$

3. Find the final solution using the third and subsequent steps of the algorithm presented in Section 2.1 using the following values of $\check{b}$ and $\hat{b}$

$$\check{b} = \underline{\check{b}} + \triangle^* \tag{25}$$

$$\hat{b} = \underline{\hat{b}} + \nabla^* \tag{26}$$

Remark 2. (About $c^\triangle$ and c^∇) In this method, we defined $c^\triangle$ and c^∇ as weights of $\triangle$ and ∇. In this chapter, we use $c_i^\triangle$ and c_i^∇ as the unitary cost associated to increase each resource $\underline{\check{b}}_i$ and $\underline{\hat{b}}_i$ respectively.

Therefore, $\triangle$ and ∇ are auxiliary variables that operate as a Type-reducers[2], this means that for each uncertain $\tilde{b}_i$, we obtain a fuzzy set embedded on its FOU where $\triangle_i^*$ and ∇_i^* become $\check{b}_i$ and $\hat{b}_i$ to be used in the Zimmermann's method (See Section 2.1).

6 Application Example

The proposed method is illustrated using an example where the perception of the experts of the system is used to define the constraints of the problem.

Now, if different experts provide an opinion based on their previous knowledge, the problem is about how to use the information they have provided. Sometimes, the experts use words instead of numbers to define a constraint (the demand of a product, for instance), by using sentences as *"I think that the availability of a resource b should be between b_1 and b_2"*, where b_1 and b_2 become $\check{b}_i$ and $\hat{b}_i$, as presented in Section 2.1.

When different experts have different opinions using the same words, then linguistic uncertainty appears (through $\triangle$ and ∇), and Type-2 fuzzy sets arise as a tool to handle this kind of uncertainty. In this way, we present the constraints defined by the experts, where the main idea is to maximize the profits of the system, so we need to compute $\tilde{z}$ and $z^* = c(x^*)$ using $\tilde{b}$, c and A which are provided as follows.

$$A = \begin{bmatrix} 3 & 3 & 4 & 5 \\ 2 & 3 & 4 & 8 \\ 6 & 7 & 4 & 2 \\ 5 & 1 & 3 & 2 \\ 3 & 2 & 3 & 2 \end{bmatrix} \quad \underline{\check{b}}_i = \begin{bmatrix} 20 \\ 30 \\ 25 \\ 35 \\ 20 \end{bmatrix} \quad \bar{\check{b}}_i = \begin{bmatrix} 40 \\ 35 \\ 30 \\ 40 \\ 35 \end{bmatrix} \quad \underline{\hat{b}}_i = \begin{bmatrix} 45 \\ 40 \\ 40 \\ 45 \\ 30 \end{bmatrix} \quad \bar{\hat{b}}_i = \begin{bmatrix} 60 \\ 45 \\ 50 \\ 55 \\ 50 \end{bmatrix}$$

$$c_{ij} = \begin{bmatrix} 7 \\ 5 \\ 7 \\ 9 \end{bmatrix} \quad c^\triangle = \begin{bmatrix} 0.5 \\ 0.2 \\ 0.5 \\ 0.2 \\ 0.5 \end{bmatrix} \quad \triangle = \begin{bmatrix} 20 \\ 5 \\ 5 \\ 5 \\ 15 \end{bmatrix} \quad c^\nabla = \begin{bmatrix} 0.5 \\ 0.5 \\ 0.2 \\ 0.2 \\ 0.5 \end{bmatrix} \quad \nabla = \begin{bmatrix} 15 \\ 5 \\ 10 \\ 10 \\ 20 \end{bmatrix}$$

---

[2] A Type-reduction strategy regards to a method for finding a single fuzzy set embedded into the FOU of a Type-2 fuzzy set.

This example is composed by four variables and five constraints whose parameters are defined by experts using IT2FS, so we apply the Algorithm shown in Section 5 to find a solution of the problem. The obtained fuzzy set $\tilde{Z}$ is defined by the following boundaries:

$$\bar{\bar{z}} = 41.667$$
$$\hat{\bar{z}} = 70.9091$$
$$\underline{\bar{z}} = 57.7273$$
$$\hat{\underline{z}} = 84.0909$$

6.1   Obtained Results

After applying the LP models shown in (22) and (24), the values of $\check{z}^*$ and $\hat{z}^*$ are 49.9091 and 79.5909 respectively. By applying the Zimmermann's method we obtain a crisp solution of $\alpha^* = 0.6099$ and $z^* = 68.012$. A detailed report of the obtained solution is shown next.

$$
\begin{aligned}
&\triangle_1^* = 0 && \nabla_1^* = 8.6364 && x_1^* = 51.049 \\
&\triangle_2^* = 5 && \nabla_2^* = 5 && x_2^* = 0 \\
&\triangle_3^* = 10 && \nabla_3^* = 5 && x_3^* = 0 \\
&\triangle_4^* = 0 && \nabla_4^* = 0 && x_4^* = 35.864 \\
&\triangle_5^* = 0 && \nabla_5^* = 0 &&
\end{aligned}
$$

The optimal solution is provided in terms of x_j^* which are the optimal decision variables, regarding the optimal satisfaction degree α. Figure 6 shows the Type-reduced fuzzy set of optimal solutions $\check{z}$ which is embedded into the FOU of $\tilde{Z}$ (See Figueroa and Hernández [28]), where the global satisfaction degree of $\alpha^* = 0.6099$ allow us to find a solution of the problem, which leads to the above values of x_j^*.

6.2   Discussion of the Results

As expected, an optimal solution of the problem is obtained in terms of x^* and α^*. For the sake of understanding, the proposed method obtains a fuzzy set embedded into the FOU of $\tilde{b}$ and $\tilde{Z}$; this set is then used by Zimmermann's method which finally returns the values of x^* and α^*.

The obtained $\check{z}$ and $\hat{z}$ come from $c^\triangle$ and c^∇, so the method selects only the auxiliary variables which improve the goal of the system. Note that even when the method incurs in additional costs for using $\triangle$ and ∇, the global solution is improved because $c^\triangle$ and c^∇ were absorbed by the reduced costs of the model.

This happens because the method selects the constraints that increase the objective function, accomplishing (17) instead of the natural reasoning of treating all constraints in the same way (using either proportional or linear increments to find a combination of $\check{z}$ and $\hat{z}$), due to it uses $c^\triangle$ and c^∇.

The analyst faces the problem of having an infinite amount of possible choices of x_j, so we point out that our approach helps decision making, based on a Type-reduction strategy to reduce the complexity of the problem, getting results which

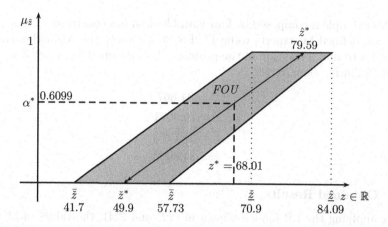

Fig. 6. Interval Type-2 fuzzy set $\tilde{z}$ embedded into the FOU of $\tilde{Z}$

are a selection made from the possible set of choices embedded into $\triangle$ and ∇, using α^* as defuzzifier.

7   Concluding Remarks

The proposed method is able to deal with Type-2 fuzzy constraints using well known fuzzy optimization techniques, achieving a solution of the problem.

Figure 4 shows the proposed methodology for designing optimization procedures for IT2FLP problems. Different Type-reduction strategies may be used, so the reader can use our methodology as a tool for new results.

The proposed method works alongside with the Zimmermann's method for finding a solution to a Type-2 constrained problem, using LP methods which have high interpretability. This means that the problem can be solved using well known algorithms, with a high interpretability and applicability of their results.

Finally, the proposed methodology is intended to be a guide about how to address a problem which includes Type-2 fuzzy constraints, involving the opinions and perceptions of different experts, using their previous knowledge and non-probabilistic uncertainty. Other methods can be applied, so our proposal is only an approach to solve this kind of problems.

References

1. Ghodousiana, A., Khorram, E.: Solving a linear programming problem with the convex combination of the max-min and the max-average fuzzy relation equations. Applied Mathematics and Computation 180(1), 411–418 (2006)
2. Sy-Ming, G., Yan-Kuen, W.: Minimizing a linear objective function with fuzzy relation equation constraints. Fuzzy Optimization and Decision Making 1(4), 347 (2002)

3. Tanaka, H., Asai, K.: Fuzzy Solution in Fuzzy Linear Programming Problems. IEEE Transactions on Systems, Man and Cybernetics 14, 325–328 (1984)
4. Tanaka, H., Asai, K., Okuda, T.: On Fuzzy Mathematical Programming. Journal of Cybernetics 3, 37–46 (1974)
5. Inuiguchi, M., Sakawa, M.: A possibilistic linear program is equivalent to a stochastic linear program in a special case. Fuzzy Sets and Systems 76(1), 309–317 (1995)
6. Inuiguchi, M., Sakawa, M.: Possible and necessary optimality tests in possibilistic linear programming problems. Fuzzy Sets and Systems 67, 29–46 (1994)
7. Inuiguchi, M., Ramík, J.: Possibilistic linear programming: a brief review of fuzzy mathematical programming and a comparison with stochastic programming in portfolio selection problem. Fuzzy Sets and Systems 111, 3–28 (2000)
8. Angelov, P.P.: Optimization in an intuitionistic fuzzy environment. Fuzzy Sets and Systems 86(3), 299–306 (1997)
9. Dubey, D., Chandra, S., Mehra, A.: Fuzzy linear programming under interval uncertainty based on ifs representation. Fuzzy Sets and Systems 188(1), 68–87 (2012)
10. Zimmermann, H.J.: Fuzzy programming and Linear Programming with several objective functions. Fuzzy Sets and Systems 1(1), 45–55 (1978)
11. Zimmermann, H.J., Fullér, R.: Fuzzy Reasoning for solving fuzzy Mathematical Programming Problems. Fuzzy Sets and Systems 60(1), 121–133 (1993)
12. Dantzig, G.: Linear Programming and Extensions. Princeton (1998)
13. Bazaraa, M.S., Jarvis, J.J., Sherali, H.D.: Linear Programming and Networks Flow. John Wiley and Sons (2009)
14. Bellman, R.E., Zadeh, L.A.: Decision-making in a fuzzy environment. Management Science 17(1), 141–164 (1970)
15. Kall, P., Mayer, J.: Stochastic Linear Programming: Models, Theory, and Computation. Springer (2010)
16. Mora, H.M.: Optimización no lineal y dinámica. Universidad Nacional de Colombia (2001)
17. Mendel, J.: Uncertain Rule-Based Fuzzy Logic Systems: Introduction and New Directions. Prentice Hall (2001)
18. Mendel, J., John, R.I., Liu, F.: Interval type-2 fuzzy logic systems made simple. IEEE Transactions on Fuzzy Systems 14(6), 808–821 (2006)
19. Mendel, J., John, R.I.: Type-2 fuzzy sets made simple. IEEE Transactions on Fuzzy Systems 10(2), 117–127 (2002)
20. Liang, Q., Mendel, J.: Interval type-2 fuzzy logic systems: Theory and design. IEEE Transactions on Fuzzy Systems 8(5), 535–550 (2000)
21. Karnik, N.N., Mendel, J.: Operations on type-2 fuzzy sets. Fuzzy Sets and Systems 122, 327–348 (2001)
22. Karnik, N.N., Mendel, J., Liang, Q.: Type-2 fuzzy logic systems. Fuzzy Sets and Systems 17(10), 643–658 (1999)
23. Melgarejo, M.: A Fast Recursive Method to compute the Generalized Centroid of an Interval Type-2 Fuzzy Set. In: Annual Meeting of the North American Fuzzy Information Processing Society (NAFIPS), pp. 190–194. IEEE (2007)
24. Melgarejo, M.: Implementing Interval Type-2 Fuzzy processors. IEEE Computational Intelligence Magazine 2(1), 63–71 (2007)
25. Figueroa, J.C.: Linear programming with interval type-2 fuzzy right hand side parameters. In: 2008 Annual Meeting of the IEEE North American Fuzzy Information Processing Society, NAFIPS (2008)
26. Figueroa, J.C.: Solving fuzzy linear programming problems with interval type-2 RHS. In: 2009 Conference on Systems, Man and Cybernetics, pp. 1–6. IEEE (2009)

27. Figueroa, J.C.: Interval type-2 fuzzy linear programming: Uncertain constraints. In: IEEE Symposium Series on Computational Intelligence, pp. 1–6. IEEE (2011)
28. Figueroa-García, J.C., Hernandez, G.: Computing optimal solutions of a linear programming problem with interval type-2 fuzzy constraints. In: Corchado, E., Snášel, V., Abraham, A., Woźniak, M., Graña, M., Cho, S.-B. (eds.) HAIS 2012, Part I. LNCS, vol. 7208, pp. 567–576. Springer, Heidelberg (2012)
29. Mendel, J.: Fuzzy sets for words: a new beginning. In: The IEEE International Conference on Fuzzy Systems, pp. 37–42 (2003)
30. Mendel, J.: Type-2 Fuzzy Sets: Some Questions and Answers. IEEE coNNectionS. A Publication of the IEEE Neural Networks Society (8), 10–13 (2003)
31. Mendel, J.M., Liu, F.: Super-exponential convergence of the Karnik-Mendel algorithms for computing the centroid of an interval type-2 fuzzy set. IEEE Transactions on Fuzzy Systems 15(2), 309–320 (2007)
32. Kearfott, R.B., Kreinovich, V.: Beyond convex? global optimization is feasible only for convex objective functions: A theorem. Journal of Global Optimization 33(4), 617–624 (2005)
33. Niewiadomski, A.: On Type-2 fuzzy logic and linguistic summarization of databases. Bulletin of the Section of Logic 38(3), 215–227 (2009)
34. Niewiadomski, A.: Imprecision measures for Type-2 fuzzy sets: Applications to linguistic summarization of databases. In: Rutkowski, L., Tadeusiewicz, R., Zadeh, L.A., Zurada, J.M. (eds.) ICAISC 2008. LNCS (LNAI), vol. 5097, pp. 285–294. Springer, Heidelberg (2008)
35. Wolsey, L.A.: Integer Programming. John Wiley and Sons (1998)
36. Papadimitriou, C.H., Steiglitz, K.: Combinatorial Optimization: Algorithms and Complexity. Dover Publications (1998)

Epistemic Considerations on Expert Disagreement, Normative Justification, and Inconsistency Regarding Multi-criteria Decision Making

Luciana Garbayo

Philosophy Department, University of Texas at El Paso,
El Paso, TX 79968, USA
lsgarbayo@utep.edu

Abstract. This paper discusses some epistemic aspects of legitimate expert disagreement between domain scientists, while considering domain specific multi-criteria decision-making problems. Particularly, it articulates both 1) the problem of the normative justification for explaining conflicting expert propositional knowledge, and also 2) the handling of disagreement derived from non-conclusive evidence, standing-in as descriptive properties of expert beliefs. Further, 3) it considers some preliminary consequences of the resulting inconsistency in the automation of conflicting expert multi-criteria decision making, and suggests that the epistemic treatment of this procedure may help to clarify what types of solution and difficulties may be there regarding the many dimensions of knowledge justification.

1 Epistemology, Propositional Knowledge and the Pivotal Role of Epistemic Justification in Disagreement

Epistemology or theory of knowledge is the special branch of philosophy that carefully investigates conceptually what knowledge *is*. With Bertrand Russell's critical re-consideration of both scientific and ordinary knowledge in an analytic key, in the turn of the 20th century [19], the field of epistemology passed to focus especially on the study of *propositional* knowledge as knowledge of facts, or descriptive knowledge, enunciated in clear declarative sentences. To know – propositionally – is to know *that* something is the case (as opposed to knowing how, or simply, knowing "intuitively"). To know *that* something is the case depends on the ability of the knower to support such declarative content with reasons and/or evidence. Hence, a claim of propositional knowledge is neither self-evident, nor private – it gives itself to public scrutiny and debate, despite being eventually held individually.

The traditional analysis of such type of propositional knowledge demanding of reasons and/or evidence is given in the *justified true belief* tradition (JTB).

M. Ceberio and V. Kreinovich (eds.), *Constraint Programming and Decision Making*,   35
Studies in Computational Intelligence 539,
DOI: 10.1007/978-3-319-04280-0_5, © Springer International Publishing Switzerland 2014

In this tradition, the study of epistemic justification – the conditions necessary and sufficient for a person be said to be justified in her belief to hold true – is pivotal for understanding both agreement and disagreement among peers on holding propositional knowledge to be justified. Such concept of justification is normative (or evaluative), for it refers to an evaluation of the criteria and/or method used in approaching propositions, while seeking for a strong degree of belief. Further, the study of justification contributes to the investigation of how experts evaluate reasons and/or evidence to uphold knowledge and find eventually conflicting justifications to the revision of propositions in a shared domain. Such variability may generate multi-criteria approaches to evaluate evidence, and influence decision-making[1]. Epistemology has a role in untangling such reasons and/or evidence methodologically and thereby, in finding fruitful ways to re-describe disagreement, aiming at clarifying it productively.

2   The Justified True Belief Tradition

In order to provide a conceptual understanding of propositional knowledge, the justified true belief (JTB) tradition analyzes it in three components. Propositional knowledge requires: 1) a *belief* (as a truth-bearer) *that p* (the object of the proposition, say, "that the sky is blue"). It also requires such proposition to be 2) true of the *world* (meaning that it agrees with the truth-maker, semantically[2], referentially, etc), and 3) that such true belief *that the sky is blue* is further epistemically *justified* by the knower. To be able to fulfill a justification condition here means that a person has reached necessary and sufficient sustaining reasons and/or evidence to affirm *that p*, based on the strength of her means to access the truth of the proposition, both cognitively and scientifically.

Definition 1. (Knowledge as JTB) *A subject S knows that a proposition P is true if and only if:*

1. *P is true*
2. *S believes that P is true, and*
3. *S is justified in believing that P is true.*

---

[1] In this chapter we will not focus on the ethical aspects of decision-making and justification in science – the focus here is restricted or bracketed, to normative scientific justification to propositional knowledge descriptions – discounting other relevant societal dimensions at this time.

[2] Tarski's leading semantic approach to truth involves the correspondence relation between words and things, expressed in a T bi-conditional structure such as

"Snow is white" is true iff snow is white.

In this "T-Scheme" the same sentence goes in both parts: one as words, and the other as thing/world; see [2], Chapter 2

In the JTB tradition, other ordinary types of knowledge (knowledge by acquaintance, know-how) are mostly handled as playing a part of the justification task. They answer to the question on the justification of propositional knowledge, as it becomes extended to inquiries on the forces of contingent and necessity (*a priori* and *a posteriori*) judgments in science, on the role of perception, on the social and on the linguistic dimensions of knowledge, etc. – to name a few. In this sense, propositional knowledge is instead a second-order type of vetted and justified type of knowledge, in need of clearance over time.

Considerations on interpreting such knowledge correctly – expertly – rests also on interlocutors agreeing on the logic and context of use thereof, and just then, on its possibly different justifications[3]. Further nuances emerges when experts consider modeling the different possibilities, including the truth-functional value (many-valued, bi-valued) of a proposition, depending on the provided interpreted semantics and the logic that expresses it, to consider also positing measurement alternatives, degrees, quantification strategies, etc. A rigorous account of propositional knowledge strongly depends on considering precisely both its accurate description (with its implicit logic and semantic background assumptions) and its justification. Agreement or disagreement has to first be established by considering such structural logic-linguistic elements for interpreting propositional knowledge.

3 Justified True Belief Fails: From Knowledge to Justification, to Belief Revision

The JTB tradition has been famously attacked, in its pretension to offer an analysis of knowledge as justified true belief, given that there might be cases of justified true belief that are not instances of knowledge. Those are very

---

[3] A simple example: for instance, if someone states propositionally that: "I know *that* London is in Europe", we have to first clarify ambiguities such as "which London are you referring to?" (not a girl, or a company called London, but the city in England – and not the one in Canada). After fixing the reference London-England, semantically, we may ask how the person knows it, as a question of epistemic justification. For instance: to *know* London *tout court*, as of having been there herself, as opposed to having read about it in an authoritative book, or inferred about it from other sources of information. In both cases, it is the epistemic justification that changes – knowing London (the city) as a personal, direct experience, and knowing about London, through study, yields a very different scope of knowledge claims, with perhaps very little overlapping.

interesting cases[4] to learn from Gettier's classic work [10]. Gettier pointed out[5] that we may be justified in believing true propositions in ambiguous contexts with coincidental and accidental justification, for example, or, be lucky to hold a true proposition when there is inconclusive evidence that may agree with different interpretations we happen to hold. One may hold a JTB but not really know conclusively, yet still be right – if the justification given to such true belief depends mostly on the introspective awareness of what justifies the belief. This is the case because cognitive accessibility is a necessary and sufficient condition for knowing – what is called "internalist justification"[6].

Yet, the discrepancy learned from Gettier-like cases introduces the idea that cognitive accessibility and instrospection is not sufficient for attaining knowledge – an externalist[7], truth-conducive process[8], even if opaque to introspection, is needed – cognitive accessibility is neither necessary nor sufficient for knowledge (for instance, children and animals know, without awareness). The externalist thesis is that what matters is that evidence is associated with a true belief, even if such association is in excess of the limited awareness of any individual knower.

---

[4] For instance: "Case I: Suppose that Smith and Jones have applied for a certain job. And suppose that Smith has strong evidence for the following conjunctive proposition: (d) Jones is the man who will get the job, and Jones has ten coins in his pocket. Smith's evidence for (d) might be that the president of the company assured him that Jones would in the end be selected, and that he, Smith, had counted the coins in Jones's pocket ten minutes ago. Proposition (d) entails: (e) The man who will get the job has ten coins in his pocket. Let us suppose that Smith sees the entailment from (d) to (e), and accepts (e) on the grounds of (d), for which he has strong evidence. In this case, Smith is clearly justified in believing that (e) is true. But imagine, further, that unknown to Smith, he himself, not Jones, will get the job. And, also, unknown to Smith, he himself has ten coins in his pocket. Proposition (e) is then true, though proposition (d), from which Smith inferred (e), is false. In our example, then, all of the following are true: (i) (e) is true, (ii) Smith believes that (e) is true, and (iii) Smith is justified in believing that (e) is true. But it is equally clear that Smith does not know that (e) is true; for (e) is true in virtue of the number of coins in Smith's pocket, while Smith does not know how many coins are in Smith's pocket, and bases his belief in (e) on a count of the coins in Jones's pocket, whom he falsely believes to be the man who will get the job" ([10], p. 122 *Analysis*).

[5] Russell, in fact, has called attention to such problems before Gettier, in [19].

[6] **Definition 2.** (Internalism) *S believes b iff that which justifies b is cognitively accessible to S.*

[7] Notably, Alvin Goldman's causal theory & reliabilist epistemology: "My proposal is this. The justificational status of a belief is a function of the reliability of a process or processes that cause it, where (as a first approximation) reliability consists in the tendency of a process to produce beliefs that are true rather than false" ([11], p. 137).

[8] **Definition 3.** (Externalism) *S believes b iff such belief is justified by a truth-conducive process.*

For our purposes, the interest in the revision of the traditional internalist approach to justification and the introduction of the externalist approach is to consider their association to the problem of the normative epistemic justification in the sciences, and their role in expert disagreement. A focus on both an externalist and an internalist approaches to epistemic justification jointly would bring a way to illuminate how peers may disagree, while recognizing the same evidence, and how expertise calibration, via direct and indirect evaluation of the authority of peers on each others' authority in the subject matter, would play a role [14].

4   Fallibilism in Science, JTB and Expert Disagreements

Scientific knowledge emerges from a non-monotonic process, provided that one holds some epistemic justification aligned with the world studied, to run such iterations – even if the justifications themselves will need to be updated over time. Perhaps one of the most important lessons is that the JTB in a post-Gettier epistemology may offer a way into the modeling the dynamics of belief change and its pitfalls, while considering the weight of its different modes of justification, unequal access to evidence, etc, equated with failed or incomplete instances of knowing. Such revision is compatible with fallibilism, arguably the mainstream doctrine adopted by scientists. Fallibilism which suggests, in its epistemic dimension, that provisional propositional knowledge iterations emerge from a belief revision process – while the access to evidence and to the formulation of theory which illuminates the proposition/hypothesis and confirms it, changes over time[9], mostly, with some level of cognitive progress. In other words, knowledge, described here as JTB, in a fallibilist key, has, at most, the role of a regulative ideal, associated to a goal-oriented pursuit, in considering the cumulative dimension of evidence, and its justification.

If the characterization above is acceptable, then scientists should probably try to better understand what epistemic justification is entailed in their scientific judgments and in their disagreements. Also researchers who study such experts's activities should follow suit. In fact, recently, a new area of philosophical research has emerged within the broad field of epistemology, specifically dedicated to the study of the *epistemology of disagreement* [4]. Such new investigation niche, we believe, has a special significance for scientists, who are systematically involved in a number of disagreements of a special sort – namely, *expert disagreements*. The epistemic study of those disagreements, while considering variable knowledge standards and some degree of methodological disunity within a domain community, offers new ways to reflect upon its consequences, particularly in managing legitimate disagreements in a choice context. Here a strong argument may be made that philosophy – and epistemology, in particular

---

[9] Fallibilism is commonly accepted among the natural scientists. Further, American pragmatists, such as John Dewey and Peirce, were strong proponents of fallibilism in philosophy; Karl Popper, famously, defended that such a position is the case in science at large [18].

– can offer an important contribution to the interdisciplinary study of expert disagreement, by helping to clarify structurally the underpinnings of knowledge claims and its different types of justification.

In epistemic language, we can say that scientists *qua* experts, are, first and foremost, *epistemic peers*. As epistemic peers, they have in principle equal possibilities to get the truth-functional value of their claims right [9], given that they have roughly equal access to the same evidence, arguments, and display roughly the same basic epistemic virtues needed in the profession, such as intellectual honesty. As domain experts, they are roughly, equally prepared to evaluate propositions in their field. Disagreement then may appear in two forms: as a merely, *verbal* disagreement, when experts fail to commensurate the language that they use among themselves, and misunderstand each other (such case is supposed to happen more in interdisciplinary scenarios, where a common vocabulary has to be agreed upon for any fruitful collaboration – which is virtually ruled out in this context, among domain-specific experts). Also, it may be a case of legitimate or genuine disagreement, when, even with the same access to evidence and arguments, they may disagree on the interpretation thereof. They may then rightfully agree to disagree[10].

Experts ideally possess appropriate justification to the knowledge they have in their domain science. What is not known, can be treated expertly as in quantifying uncertainty or in establishing degrees of knowledge, and be introduced in the domain. Crucially, the awareness of knowing explicitly what you know and distinguishing what you know that you do not know (as opposed to what you do not know that you do not know), adds an important self-referential dimension to managing one's domain knowledge as an expert.

Now, epistemic justification is embedded into scientific justification: experts may probe their scientific knowledge of x through experiment, observation or modeling which ultimately, depend on justifying her beliefs epistemically – by relying on the testimony of the senses, the use of reason, the consideration of standards of coherence, of the role of background assumptions, etc. The crux of the problem of offering an expert response and the debate that ensues, resides on the strength of the justification, broadly conceived, to the claims made. It might thus be safe then to affirm that *the greater the expert, the better the epistemic justification* provided to the propositional knowledge affirmed in one's domain.

Here, two dimensions of expert disagreement in the same domain are prominent: one has to do with disagreement about descriptions of phenomena or objects, and the other has to do with disagreement about the criteria to normatively justify knowing them. In the first case, there may be expert disagreement derived from non-conclusive evidence, standing-in as descriptive properties of expert beliefs in reference to the phenomena studied in their domain. In other words,

---

[10] Here we stand against Aumann's classic position [1], who suggested that experts would convert to find agreement among them over time, given that they share the same priors – if they were Bayesian rational. We agree with Kelly [13], who suggests that experts may share evidence, but that the decision on which parts of it may constitute a prior, is open to peer disagreement.

experts may have reasons to disagree on the very description of their objects of study, and pick up different parts of phenomena to be representative to modeling. This fact is well known and recognized both longitudinally, in the history of science, driven by different same domain experts and rival re-descriptions of their objects ("ontology driven" [3]) over time, but also synchronically, as different scientists may provide different descriptions to the same phenomena, also given to their alliances to different theoretical ways to illuminate their objects.

Expert theoretical choices also provide, in fact, important epistemic justification to their description of phenomena. The theoretical disagreement that matters in this case is the disagreement on the criteria by which experts pick up phenomena meaningfully. Such is an evaluative criterion which illuminates the descriptive elements. In this sense, justification is normative and, by the same token, disagreement is not referenced in our ability to simply naively see something, but on the normative reasons to select from what we see, a description that can be justified. When an epistemic agent is learning to *see* a microscopic image or an X-ray, she is really not "seeing them" but selecting the phenomena from the image and evaluating it as such selection, contrasted with multiple other possibilities. She does it while considering background assumptions, methodologies, artifacts and so on.

The *evaluative or normative* justificational dimension of expert disagreement among multiple experts refers to the disagreement on the set of criteria relevant to consider the dependence relation of normative justification on the criteria of description of phenomena. Such position of normative justification seems to be the acceptance of a kind of *methodism* in epistemology [5]. Methodism is the position in epistemology that asks "how do we know" primarily, before making assertions about what is known – it requires a set of criteria. In fact, experts first have to command the criteria of what constitute knowledge in their domain fields in order to evaluate what to describe and to what extent we see it. In addition, they have to implicitly approve the processes by which we acquire evidence, such as perceptually, inferentially, etc. In this sense, epistemic justification is a positive evaluation of both the process by which one sees – perceptually and with reason, given a trained selective judgment of evidence, as a scientific justification, in connection with an externalist resource to the truth-conduciveness of the processes of picking-up phenomena.

Given that we are considering scientists in their individual spheres, we then should add that, the normative, propositional justification should be acceptable and coupled with one's doxastic justification, except that its normative dimension might be eventually not shared. As Kvanvig simply puts it ([15], p. 7):

> "Doxastic justification is what you get when you believe something for which you have propositional justification, and you base your belief on that which propositionally justifies it."

If there is propositional justification for believing, but one's normative criteria might differ in considering such reasons and/or evidence, a scientist, as an expert, may understand the position, but not give doxastic justification to it. Agreement is given when scientists assent individually, with their doxastic justification. In the case of disagreement, there is recognition of propositional justification, but no doxastic assent.

5   Dealing with Multi-criteria Decision-Making in the Context of Expert Disagreement

If methodism seems to be the case for describing the normative justification of experts who may disagree on how they know the phenomena they pick up, and do not offer doxastic justification for each other's position, then it does happen that, in an open field of research, there may be corresponding multiple expert criteria to decide what to actually pick up as the relevant phenomena. Thus, methodism accomodates a possible multi-criteria decision making problem in modeling strong disagreement on normative justification to be considered by epistemic peers, as they evolve and refine the knowledge of their domain science.

Surely, in order to identify such disagreement on multi-criteria decision making regarding normative justification, epistemic peers have to deal away first with their mere verbal disagreements, as mentioned before. Such type of disagreement is the one based on mismatched communication on equivalent standards of description, and commensurable conceptual frames. In order to overcome such distracting scenarios, experts normalize their language, control carefully their working definitions and vocabulary in general. A task of great difficulty and quite relevant to disentangle expert disagreement from its verbal difficulties, is that experts as epistemic peers should also strive to square away the role of context-sensitivity in their descriptions and the use of their criteria, for the sake of clarity. Careful preliminary considerations on context may allow them to eventually know when to correctly agree to disagree, discounting context. Only when both language is normalized and contexts are squared, we can consider the possibility of legitimate expert disagreement to be the case among experts. Just in this case, the normative justification for knowledge may finally and clearly appear prominently as a legitimate reason for contention.

Legitimate disagreement on the normative justification we suggest, can be mostly related to the methodological disunity found in philosophy and in many crucial scientific areas, whereas multiple criteria are available to evaluate and justify what is the case. Historically, such disunity becomes especially clear when the development of a consensual metrics is needed.

If one considers all experts in principle credible and takes prima facie the instances of their disagreements that are deemed legitimate, then the emerging question is how to operate some dimension of cognitive progress with such disagreements. A first alternative is to treat them all with the prominent *Equal Weight View* [7]. Such view suggests that all epistemic peers' judgments should have equal weight – provided that they have access to the same evidence, so that if they genuinely disagree, then skepticism should ensue for all positions, until further evidence is presented. An alternative and competing view is the *Total Evidence View* [13], in which it is prescribed that none of the peers give up their positions while facing disagreement, but that they keep instead firmly attached to their own views, while considering the total evidence, inclusive of the disagreement as a data point as well. In a quick analysis, we can see that in the first case, legitimate disagreement is taken to be only treatable by the emergence of a new consensus, so agnosticism should be the rule until then. But, in the second view, dissensus might be the case, and should be dealt with a different attitude, of ownership.

In further analysis, this last view seems to express most of what happens in real, legitimate, expert disagreement. Agnosticism does not seem to play a strong role in argumentation; rather, it might stale it, if not properly measured. In the case of experts standing their ground on different views, given the fact that divergent normative justification may indeed justify different evaluative criteria for knowledge, we can consider that it is more common the persistence of disagreement.

Indeed, modeling such persistence of expert disagreement is of great importance for the study of its characteristics, consequences, and its evolution over time. In this sense, rather than focusing on building an ideal model for expert consensus, here we suggest that the modeling of dissensus, from the point of view of the argumentation process regarding the consideration of normative justification for scientific claims should be encouraged. The idealized model of consensus building has of course its relevant place, but it is insufficient, for its limited ability to describe good enough real world processes. In comparison, to express legitimate expert disagreement with multi-criteria normative justification is quite relevant, so that we can better understand the totality of expert claims and the depth and many layers of disagreements, inclusive of internalist and externalist types of justification.

We suggest extending the *Total Evidence View*, to consider the combined points of view of all experts participants in a legitimate disagreement, with the degrees of expected inconsistency to be described, in the commensuration of claims and justifications. Modeling such inconsistency may add much clarification can be made in understanding types and trends in a debate among peers.

6   Computer Science and Epistemology of Disagreement: Some Initial Convergent Notes

In solidarity with philosophy, computer scientists developed strategies to deal with such problems of modeling disagreement by constituting a constraint semantics to solve combinatorial search problems, such as considering optimizing the inconsistency of expert argumentation, as a soft constraint satisfaction problem. In particular, Dung's theory of argumentation (AF), designed to graph sets of arguments considering binary conflicts based on attack relations between them. In their description, the resulting arguments are weighted in AF based on fuzziness, probability, and preference [6]. As a preliminary response, it seems that such description in the case of the disagreement of experts is incomplete. Here we focus on the relevance of epistemology in adding important new categories to classify what disagreement is about, in its self-referential features as controlling factors, and to better understand and analyze the inconsistency of the whole. The case in point is to describe normative justification for constraint semantics, in order to have a better dynamic picture of what happens in legitimate expert disagreement.

In this sense, we preliminarily suggest that the epistemic treatment of the automation procedure of multi-criteria decision making be a necessary layer of information, so that disagreement is not clearly not only verbal, or ordinary, but captures what is special about it, in the case of its legitimacy, including its normative justification, the consideration of both internalist and externalist strategies, and the assent of doxastic justification. The epistemic dimension as just described, may be indeed part of the missing ontology of AF, given that it is based only on "Beliefs and goals" [6], and not on claims and justifications, that could be described in terms of justified true beliefs. The justificational dimension, we would like to suggest, is the benchmark for reconsidering productively the inconsistency in the whole of argumentation in the expert case, while experts are redefining the field, as they revise normativity justificational standards proper in their area – rather than lumping it with mere non-expert opinion, which are mainly distractions.

7   Conclusion and Future Work: On Modeling Methodological Disunity in Legitimate Disagreement Contexts

The lessons quasi-learned so far, we suggest, are that, if we accept legitimate disagreement to be the case among experts, we should first consider the normativity embedded in the justifications, and be able to model it accordingly, with its nuances and degrees. Part of it, may be resolved with simulation analytics [20], mitigating ambiguities and introducing projections as a way to generate further consideration of justification. But, further, to express such deep disagreements and all its degrees, we also may need a non-classical logic approach for manipulating methodological disunity in legitimate disagreement contexts with its automation. Wang, Ceberio, et al. [21] explore fuzzy logic as an alternative to characterize disagreement among peers, and the problems of multi-criteria decision-making associated with such disagreements in a computational model. We would like to suggest, with Martine Ceberio (oral communication), that disagreement put us on the path of studying non-classic logic in general to describe dissensus, and may contribute for a methodic way to model what. There is exciting work ahead to be done, in a equally non-monotonic, self-revising way.

References

1. Aumann, R.J.: Agreeing to Disagree. The Annals of Statistics 4(6), 1236–1239 (1976)
2. Burgess, A.G., Burgess, G.P.: Truth. Princeton University Press, Princeton (2011)

3. Cao, T.Y.: Conceptual Developments of Twentieth Century Field Theories. Cambridge University Press, Cambridge (1997)
4. Christensen, D.: Special Issue: "The Epistemology of Disagreement". Episteme: A Journal of Social Epistemology 6(3), 231–353 (2009)
5. Chisholm, R.: Theory of knowledge. Prentice Hall, Englewood Cliffs (1989)
6. Dung, F.M.: On the acceptability of arguments and its fundamental role in non-monotonic reasoning, logic programming and n-person games. Artificial Intelligence 77(2), 321–358 (1995)
7. Feldman, R.: Epistemology. Prentice Hall, Englewood Cliffs (2003)
8. Feldman, R., Warfield, T.A. (eds.): Disagreement. Oxford University Press, Oxford (2010)
9. Foley, R.: Intellectual trust in Oneself and Others. Cambridge University Press, Cambridge (2001)
10. Gettier, E.: Is Justified True Belief Knowledge? Analysis 23, 121–123 (1963)
11. Goldman, A.: What Is Justified True Belief? In: Pappas, G.R. (ed.) Justification and Knowledge. D. Reidel, Dordrecht (1979)
12. Kelly, T.: The epistemic significance of disagreement. In: Hawthorne, J., Gendler, T. (eds.) Oxford Studies in Epistemolgy, vol. 1. Oxford University Press (2005)
13. Kelly, T.: Peer Disagreement and Higher Order Evidence. In: Goldman, A.I., Whitcomb, D. (eds.) Social Epistemology: Essential Readings. Oxford University Press (2010)
14. Kitcher, P.: The Advancement of Science: Science without Legend, Objectivity without Illusion. Oxford University Press, Oxford (1993)
15. Kvanvig, J.: Propositionalism and the Perspectival Character of Justification. American Philosophical Quarterly 40(1), 3–18 (2003)
16. Lewis, D.: Elusive Knowledge. Australasian Journal of Philosophy 74, 549–567 (1996)
17. Peirce, C.S., Wiener, P.P.: Charles S. Peirce: Selected Writings. Dover, New York (1980)
18. Popper, K.: Conjectures and Refutations. Routledge, London (1963)
19. Russell, B.: The Problems of Philosophy. Cosimo Classics. Bibliobazaar Open Source Project (2007); 1st edn. (1912)
20. Stahl, J., Garbayo, L.: Simulation analytics as a tool to mitigate medical disagreements (manuscript)
21. Wang, X., Ceberio, M., Virani, S., Garcia, A., Cummins, J.: A Hybrid Algorithm to Extract Fuzzy Measures for Software Quality Assessment. Journal of Uncertain Systems 7(3), 219–237 (2013)

Interval Linear Programming Techniques in Constraint Programming and Global Optimization

Milan Hladík and Jaroslav Horáček

Charles University, Faculty of Mathematics and Physics,
Department of Applied Mathematics, Malostranské nám. 25, 118 00,
Prague, Czech Republic
{hladik,horacek}@kam.mff.cuni.cz

Abstract. We consider a constraint programming problem described by a system of nonlinear equations and inequalities; the objective is to tightly enclose all solutions. First, we linearize the constraints to get an interval linear system of equations and inequalities. Then, we adapt techniques from interval linear programming to find a polyhedral relaxation to the solution set. The linearization depends on a selection of the relaxation center; we discuss various choices and give some recommendations. The overall procedure can be iterated and thus serves as a contractor.

Keywords: Interval computation, linear programming, constraint programming, global optimization.

1 Introduction

A constraint programming problem [6, 9, 10] is usually formulated as follows. Consider equality and inequality constraints

$$f_i(x) = 0, \quad i = 1, \ldots, m, \tag{1a}$$
$$g_j(x) \leq 0, \quad j = 1, \ldots, \ell, \tag{1b}$$

or, in compact form,

$$f(x) = 0,$$
$$g(x) \leq 0,$$

where $f_i, g_j : \mathbb{R}^n \mapsto \mathbb{R}$ are real-valued functions and $f(x) = (f_1(x), \ldots, f_m(x))$, $g(x) = (g_1(x), \ldots, g_\ell(x))$. The objective is to enclose all solutions of the constraint system that lie inside a given box $x = [\underline{x}, \overline{x}]$. Similar problem is solved in global optimization, where a global minimum of a function $\varphi(x)$ subject to (1) is searched for. This makes global optimization (seemingly) more complex, but basic tools from constraint programming are intensively utilized there as well.

The fundamental idea behind our approach in solving (1) is to linearize the constraints, and then adapt interval linear programming techniques. Linear relaxations were also studied e.g. in [1, 2, 5, 20, 27], and such polyhedral relaxations

M. Ceberio and V. Kreinovich (eds.), *Constraint Programming and Decision Making,*
Studies in Computational Intelligence 539,
DOI: 10.1007/978-3-319-04280-0_6, © Springer International Publishing Switzerland 2014

were applied e.g. in solving global optimization problems [17, 19], or in control theory [25]. Interval polyhedra as a new abstract domain were investigated in [5], but their applicability for constraint programming is not well supported yet. In [1], linear programming was applied to verify infeasibility of a special satisfiability constraint program. Convex polyhedral approximation of quadratic terms in the constraints was proposed in [20]. A general polyhedral relaxation by using interval Taylor was investigated in [2, 21], and inner / outer linearizations in [26].

An n-dimensional box can be regarded as an n-by-1 interval matrix. In general, an interval matrix $\boldsymbol{A}$ is defined as

$$\boldsymbol{A} := [\underline{A}, \overline{A}] = \{A \in \mathbb{R}^{m \times n}; \ \underline{A} \le A \le \overline{A}\},$$

where $\underline{A}, \overline{A} \in \mathbb{R}^{m \times n}$ are given. The midpoint and radius of $\boldsymbol{A}$ are defined respectively as

$$A^c := \frac{1}{2}(\underline{A} + \overline{A}), \quad A^\Delta := \frac{1}{2}(\overline{A} - \underline{A}).$$

The set of all m-by-n interval matrices is denoted by $\mathbb{IR}^{m \times n}$.

Let us recall some results that we will utilize in this paper. A vector x is a solution of an interval system $\boldsymbol{A}x = \boldsymbol{b}$ if it is a solution of $Ax = b$ for some $A \in \boldsymbol{A}$ and $b \in \boldsymbol{b}$. The well-known Oettli–Prager characterization [24] (cf. [7, 13]) of the solutions to $\boldsymbol{A}x = \boldsymbol{b}$ is written by means of a nonlinear system

$$|A^c x - b^c| \le A^\Delta |x| + b^\Delta.$$

Throughout this paper, the relation $\le$ and the notion of non-negativity etc. are understood component-wise. If one knows a priori that x is non-negative, then the above reduces to the linear system of inequalities

$$\underline{A}x \le \overline{b}, \ \overline{A}x \ge \underline{b}, \ x \ge 0.$$

Similar reduction is possible when x is non-positive or lies in any orthant. For interval linear inequalities $\boldsymbol{A}x \le \boldsymbol{b}$, the description of all solutions is due to Gerlach [8] (cf. [7, 13])

$$A^c x \le A^\Delta |x| + \overline{b}.$$

Again, non-negativity of variables simplifies the nonlinear description to a linear one

$$\underline{A}x \le \overline{b}, \ x \ge 0.$$

2 Interval Linear Programming Approach

Our approach is based on linearization of constraints (1) by means of interval linear equations and inequalities; by using interval linear programming techniques [12] we construct a polyhedral enclosure to the solution set of (1) and contract the initial box $\boldsymbol{x}$. The process can be iterated, resulting in a nested sequence of boxes enclosing the solution set.

Let $x^0 \in \boldsymbol{x}$; this point will be called *the center* of linearization throughout the paper. Suppose that a vector function $h : \mathbb{R}^n \mapsto \mathbb{R}^s$ has the following linear enclosure on $\boldsymbol{x}$

$$h(x) \subseteq S_h(\boldsymbol{x}, x^0)(x - x^0) + h(x^0), \quad \forall x \in \boldsymbol{x} \tag{2}$$

for a suitable interval-valued function $S_h : \mathbb{IR}^n \times \mathbb{R}^n \mapsto \mathbb{IR}^{s \times n}$. This is traditionally calculated by a mean value form [22]. More efficiently, one can employ a successive mean value approach (as was done in [2]) or slopes; see [11, 22]. Alternatively, in some situations, the relaxation can be established by analysing the structure of $h(x)$. For example, relaxing quadratic terms can be done separately by using McCorming-like linearizations [20].

We will apply (2) for both functions f and g. Thus, the solution set to (1) is enclosed in the solution set to the interval linear system

$$S_f(\boldsymbol{x}, x^0)(x - x^0) + f(x^0) = 0, \tag{3a}$$
$$S_g(\boldsymbol{x}, x^0)(x - x^0) + g(x^0) \le 0, \tag{3b}$$

for any $x^0 \in \boldsymbol{x}$. Note that, in principle, we need not use the same x^0 in both sub-systems, however we will do so. Now, we will address the question what choice of $x^0 \in \boldsymbol{x}$ is suitable. For the sake of simplicity, denote (3) by

$$\boldsymbol{A}(x - x^0) + f(x^0) = 0, \tag{4a}$$
$$\boldsymbol{B}(x - x^0) + g(x^0) \le 0. \tag{4b}$$

2.1   Vertex Selection of x^0

Let $x^0 := \underline{x}$. Since $x - \underline{x}$ is non-negative, the Oettli–Prager theorem describes the solution set of $\boldsymbol{A}(x - \underline{x}) + f(\underline{x}) = 0$ by linear inequalities

$$\underline{A}(x - \underline{x}) + f(\underline{x}) \le 0, \quad \overline{A}(x - \underline{x}) + f(\underline{x}) \ge 0,$$

or

$$\underline{A}x \le \underline{A}\underline{x} - f(\underline{x}), \quad \overline{A}x \ge \overline{A}\underline{x} - f(\underline{x}). \tag{5}$$

By Gerlach theorem, the solution set to $\boldsymbol{B}(x - \underline{x}) + g(\underline{x}) \le 0$ is described by

$$\underline{B}(x - \underline{x}) + g(\underline{x}) \le 0,$$

or

$$\underline{B}x \le \underline{B}\underline{x} - g(\underline{x}). \tag{6}$$

Let $x^0 := \overline{x}$. Now, $x - \overline{x}$ is non-positive, the solution set to $\boldsymbol{A}(x - \overline{x}) + f(\overline{x}) = 0$ is described by

$$\overline{A}x \le \overline{A}\overline{x} - f(\overline{x}), \quad \underline{A}x \ge \underline{A}\overline{x} - f(\overline{x}), \tag{7}$$

and the solution set to $\boldsymbol{B}(x - \overline{x}) + g(\overline{x}) \le 0$ is described by

$$\overline{B}x \le \overline{B}\overline{x} - g(\overline{x}). \tag{8}$$

We can choose any other vertex of the box x and accordingly obtain a linear description; cf. [2, 26]. Employing all vertices is superfluous since there is 2^n of them. In [2], it is recommended to use two opposite corners of x; the other corners would not significantly increase efficiency. Which pair of the opposite corners is the best choice is still an open question; a random selection seems to be acceptable.

2.2   Non-vertex Selection of x^0

Let $x^0 \in x$, not necessarily a vertex; the midpoint might be a good choice. The solution set to $A(x - x^0) + f(x^0) = 0$ is described by

$$|A^c(x - x^0) + f(x^0)| \le A^\Delta |x - x^0|, \tag{9}$$

and the solution set to $B(x - x^0) + g(x^0) \le 0$ is described by

$$B^c(x - x^0) \le B^\Delta |x - x^0| - g(x^0). \tag{10}$$

These systems are not linear due to the absolute values. To get rid of them, we will linearize them by the means of Beaumont [4].

Theorem 1 (Beaumont, 1998). *Let $y \in \mathbb{IR}$ with positive radius. For every $y \in y$ one has*

$$|y| \le \alpha y + \beta, \tag{11}$$

where

$$\alpha = \frac{|\overline{y}| - |\underline{y}|}{\overline{y} - \underline{y}} \quad and \quad \beta = \frac{\overline{y}|\underline{y}| - \underline{y}|\overline{y}|}{\overline{y} - \underline{y}}.$$

Moreover, if $\underline{y} \ge 0$ or $\overline{y} \le 0$ then (11) holds as equation.

This theorem helps in linearizing any of the above absolute values. Particularly for $x^0 := x^c$, we get linearizations

$$|A^c(x - x^c) + f(x^0)| \le A^\Delta x^\Delta,$$

or,

$$A^c x \le A^c x^c + A^\Delta x^\Delta - f(x^c), \quad -A^c x \le -A^c x^c + A^\Delta x^\Delta + f(x^c),$$

for equations, and

$$B^c x \le B^c x^c + B^\Delta x^\Delta - g(x^c)$$

for inequalities.

Denote by D_v the diagonal matrix with entries $v_1, \ldots, v_p$. For an arbitrary $x^0 \in x$, linearization by Theorem 1 gives the following result.

Proposition 1. *Let $x^0 \in \boldsymbol{x}$. Then (4) has a linear relaxation*

$$(A^c - A^\Delta D_\alpha)x \le A^c x^0 + A^\Delta v^0 - f(x^0), \tag{12a}$$

$$(-A^c - A^\Delta D_\alpha)x \le -A^c x^0 + A^\Delta v^0 + f(x^0), \tag{12b}$$

$$(B^c - B^\Delta D_\alpha)x \le B^c x^0 + B^\Delta v^0 - g(x^0), \tag{12c}$$

where $\alpha_i = \frac{1}{x_i^\Delta}(x_i^c - x_i^0)$ *and* $v_i^0 = \frac{1}{x_i^\Delta}(x_i^c x_i^0 - \overline{x}_i \underline{x}_i)$.

Proof. First, we show the relaxation for (4b). By Theorem 1, (10) is relaxed as

$$B^c(x - x^0) \le B^\Delta |x - x^0| - g(x^0) \le B^\Delta(D_\alpha(x - x^0) + \beta) - g(x^0),$$

where

$$\alpha_i = \frac{1}{2x_i^\Delta}(|\overline{x}_i - x_i^0| - |\underline{x}_i - x_i^0|) = \frac{1}{2x_i^\Delta}(\overline{x}_i - x_i^0 - (x_i^0 - \underline{x}_i)),$$

$$= \frac{1}{x_i^\Delta}(x_i^c - x_i^0),$$

$$\beta_i = \frac{1}{2x_i^\Delta}((\overline{x}_i - x_i^0)|\underline{x}_i - x_i^0| - (\underline{x}_i - x_i^0)|\overline{x}_i - x_i^0|)$$

$$= \frac{1}{2x_i^\Delta}((\overline{x}_i - x_i^0)(x_i^0 - \underline{x}_i) - (\underline{x}_i - x_i^0)(\overline{x}_i - x_i^0)) = \frac{1}{x_i^\Delta}(\overline{x}_i - x_i^0)(x_i^0 - \underline{x}_i).$$

The inequality then takes the form of

$$(B^c - B^\Delta D_\alpha)x \le B^c x^0 + B^\Delta(-D_\alpha x^0 + \beta) - g(x^0).$$

Herein,

$$(-D_\alpha x^0 + \beta)_i = -\alpha_i x_i^0 + \beta_i = \frac{1}{x_i^\Delta}(-(x_i^c - x_i^0)x_i^0 + (\overline{x}_i - x_i^0)(x_i^0 - \underline{x}_i))$$

$$= \frac{1}{x_i^\Delta}(-x_i^c x_i^0 + x_i^0 x_i^0 + \overline{x}_i x_i^0 - x_i^0 x_i^0 - \overline{x}_i \underline{x}_i + x_i^0 \underline{x}_i)$$

$$= \frac{1}{x_i^\Delta}(-x_i^c x_i^0 + \overline{x}_i x_i^0 - \overline{x}_i \underline{x}_i + x_i^0 \underline{x}_i) = \frac{1}{x_i^\Delta}(x_i^c x_i^0 - \overline{x}_i \underline{x}_i) = v_i^0.$$

Now, we prove (12a)–(12b). By Theorem 1, (9) is relaxed as

$$|A^c(x - x^0) + f(x^0)| \le A^\Delta |x - x^0| \le A^\Delta(D_\alpha(x - x^0) + \beta),$$

from which

$$(A^c - A^\Delta D_\alpha)x \le A^c x^0 + A^\Delta(-D_\alpha x^0 + \beta) - f(x^0),$$

$$(-A^c - A^\Delta D_\alpha)x \le -A^c x^0 + A^\Delta(-D_\alpha x^0 + \beta) + f(x^0).$$

$\square$

2.3   Convex Case

In the proposition below, an inequality is called *a consequence* of a system of inequalities if it can be expressed as a non-negative linear combination of the inequalities in the system. In other words, it is a redundant constraint if added to the system.

Proposition 2. *Let $x^0 \in x$, but not a vertex of x. Suppose that A and B do not depend on a selection of x^0.*

1. *If $f_i(x)$, $i = 1, \ldots, m$ are convex, then the inequality (12a) is a consequence of the corresponding inequalities derived by vertices of x.*
2. *If $f_i(x)$, $i = 1, \ldots, m$ are concave, then the inequality (12b) is a consequence of the corresponding inequalities derived by vertices of x.*
3. *If $g_j(x)$, $j = 1, \ldots, \ell$ are convex, then the inequality (12c) is a consequence of the corresponding inequalities derived by vertices of x.*

Proof. We prove the item 3; the other items are proved analogously. Let $x^1, x^2 \in x$ and consider a convex combination $x^0 := \lambda x^1 + (1 - \lambda)x^2$ for any $\lambda \in [0, 1]$. It suffices to show that the inequality derived from x^0 is a convex combination of those derived from x^1 and x^2. For x^1 and x^2, the associated systems (12c) read respectively

$$(B^c - B^\Delta D_{\alpha^1})x \le B^c x^1 + B^\Delta v^1 - g(x^1), \tag{13a}$$

$$(B^c - B^\Delta D_{\alpha^2})x \le B^c x^2 + B^\Delta v^2 - g(x^2), \tag{13b}$$

where $\alpha_i^1 = \frac{1}{x_i^\Delta}(x_i^c - x_i^1)$, $\alpha_i^2 = \frac{1}{x_i^\Delta}(x_i^c - x_i^2)$, $v_i^1 = \frac{1}{x_i^\Delta}(x_i^c x_i^1 - \overline{x}_i \underline{x}_i)$, and $v_i^2 = \frac{1}{x_i^\Delta}(x_i^c x_i^2 - \overline{x}_i \underline{x}_i)$. Multiplying (13a) by λ and (13b) by $(1 - \lambda)$, and summing up, we get

$$(B^c - B^\Delta D_\alpha)x \le B^c x^0 + B^\Delta v^0 - \lambda g(x^1) - (1 - \lambda)g(x^2),$$

where $\alpha_i = \frac{1}{x_i^\Delta}(x_i^c - x_i^0)$ and $v_i^0 = \frac{1}{x_i^\Delta}(x_i^c x_i^0 - \overline{x}_i \underline{x}_i)$. By convexity of g, we derive

$$(B^c - B^\Delta D_\alpha)x \le B^c x^0 + B^\Delta v^0 - g(x^0),$$

which is the inequality (12c) corresponding to x^0.   □

The functions $f_i(x)$, $-f_i(x)$ or $g_j(x)$ need not be convex (and mostly they are not). However, if it is the case, Proposition 1 is fruitful only when x^0 is a vertex of x; otherwise, the resulting inequalities are redundant. Notice that this may not be the case for the original interval inequalities (4b).

When $f_i(x)$, $-f_i(x)$ or $g_j(x)$ are not convex, non-vertex selection of $x^0 \in \boldsymbol{x}$ may be convenient. Informally speaking, the more non-convex the functions are the more desirable may be an interior selection of x^0.

2.4 Summary

To obtain as tight polyhedral enclosure as possible it is convenient to simultaneously consider several centers for linearization. If we have no extra information, we recommend to relax according to two opposite corners of $\boldsymbol{x}$ (in agreement with [2]) and according to the midpoint $x^0 := x^c$. Putting all resulting inequalities together, we obtain a system of $3(2m + \ell)$ inequalities with respect to n variables. This system represents a convex polyhedron $\mathcal{P}$ and the intersection with $\boldsymbol{x}$ gives a new, hopefully tighter, enclosure to the solution set.

When we calculate minima and maxima in each coordinate by calling linear programming, we get a new box $\boldsymbol{x}' \subseteq \boldsymbol{x}$. Achterberg's heuristic introduced in [3] reduces the computational effort by a suitable order of solving the linear programs. Rigorous bounds on the optimal values in linear programming problems were discussed in [14, 23]. The optimal values of the linear programs are attained in at most $2n$ vertices of $\mathcal{P}$, which lie on the boundary of $\boldsymbol{x}'$. It is tempting to use some of these points as a center x^0 for the linearization process in the next iteration. Some numerical experiments have to be carried out to show how effective this idea is. Another possibility is to linearize according to these points in the current iteration and append the resulting inequalities to the description of $\mathcal{P}$. By re-optimizing the linear programs we hopefully get a tighter enclosing box $\boldsymbol{x}'$. Notice that the re-optimizing can be implemented to be very cheap. If we employ the dual simplex method to solve the linear programs and use the previous optimal solutions as starting points, then the appending of new constraints is done easily and the new optimum is found in a few steps. We append only the constraints corresponding to the current optimal solution. Thus, for each of that $2n$ linear programs, we append after its termination a system of $(2m + \ell)$ inequalities and re-optimize.

In global optimization, a lower bound of $\varphi(x)$ on $\mathcal{P}$ is computed, which updates the lower bound on the optimal value if lying in $\boldsymbol{x}$. Let x^* be a point of $\mathcal{P}$ in which the lower bound of $\varphi(x)$ on $\mathcal{P}$ is attained. Then it is promising to use x^* as a center for linearization in the next iteration. Depending on the concrete method for lower bounding of $\varphi(x)$, it may be desirable to append to $\mathcal{P}$ the inequalities (12) arising from $x^0 := x^*$, and to re-compute the lower bound of $\varphi(x)$ on the updated polyhedron.

2.5 Illustration

In the following, we give a simple symbolic illustrations of different choices of the center x^0. In the figures, $\mathcal{S}$ denotes the set described by (1), the initial box $\boldsymbol{x}$ is colored in light gray, and the linear relaxation in dark gray.

Typical situation when choosing x^0
to be a vertex:

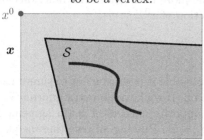

Typical situation when choosing x^0
to be the opposite vertex:

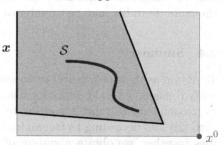

Typical situation when choosing
$x^0 = x^c$:

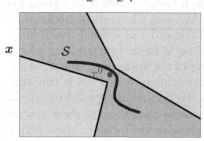

Typical situation when choosing
$x^0 = x^c$ (after linearization):

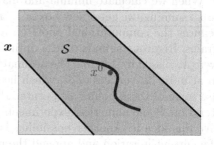

Typical situation when choosing all of them:

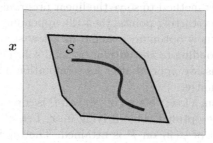

Example 1. Now, consider a concrete example with the constraints

$$\pi^2 y - 4x^2 \sin x = 0,$$
$$y - \cos\left(x + \tfrac{\pi}{2}\right) = 0.$$

where $x \in \boldsymbol{x} = [-\tfrac{\pi}{2}, \tfrac{\pi}{2}]$, and $y \in \boldsymbol{y} = [-1, 1]$.

Notice that this example can be viewed as a "hard" instance for the classical techniques because the initial box is so called 2B-consistent (the domains of variable cannot be reduced if we consider the constraints separately); see e.g. [6, 18, 20]. Also the recommended preconditioning of the system by the inverse of the Jacobian matrix for the midpoint values [11] makes almost no progress.

However, the proposed interval linear programming approach contracts significantly domains of both variables in only one iteration to $\boldsymbol{x}' = [-0.9598, 0.9598]$
and $\boldsymbol{y}' = [-0.6110, 0.6110]$.

Figures bellow illustrate the linearization for diverse selections of the center
x^0. In this example, the linearization does not depend the y-coordinate of x^0
as the derivatives of the constraint functions with respect to y are constant.
Thus, we put $x_2^0 = 0$, and varied the entry x_1^0 only. The constraint functions are
colored in red and blue. The linearized functions are depicted by gray and light
gray bands, and their intersection (which is an enclosure of the solution) in dark
gray.

The center of linearization is
$x^0 = (0,0)$.

The center of linearization is
$x^0 = (\frac{\pi}{6}, 0)$.

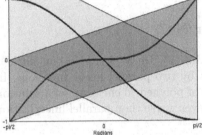

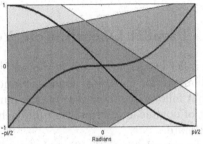

The center of linearization is
$x^0 = (\frac{\pi}{2}, 0)$.

The center of linearization is
$x^0 = (-\frac{\pi}{2}, 0)$.

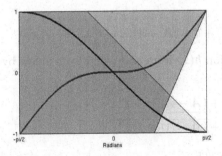

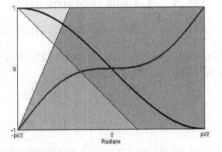

New interval enclosure after the contractions
with centers $x^0 = (0,0)$, $(\frac{\pi}{2},0)$, $(-\frac{\pi}{2},0)$.

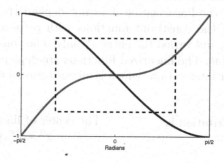

3 Parallel Linearization: Yes or No?

Parallel linearization was proposed by Jaulin [15, 16] as a simple but efficient technique for enclosing nonlinear functions by two parallel linear functions. In what follows, we show that for the purpose of polyhedral enclosure of a solution set of nonlinear systems, our approach is never worse than parallel linearization estimate.

In accordance with (2) and the subsequent, suppose that a vector function $h : \mathbb{R}^n \mapsto \mathbb{R}^s$ has the following interval linear enclosure on $\boldsymbol{x}$

$$h(x) \subseteq \boldsymbol{A}(x - x^0) + b, \quad \forall x \in \boldsymbol{x}$$

for suitable interval matrix $\boldsymbol{A} \in \mathbb{IR}^{s \times n}$ and $x^0 \in \boldsymbol{x}$, where $b := h(x^0)$.

Let $A \in \boldsymbol{A}$. Using subdistributivity

$$A(x - x^0) + b \subseteq A(x - x^0) + b + (\boldsymbol{A} - A)(x - x^0),$$

parallel linearization estimates the function $h(x)$ from above and from below by the following linear functions

$$h(x) \geq A(x - x^0) + b + \underline{(\boldsymbol{A} - A)(\boldsymbol{x} - x^0)},$$
$$h(x) \leq A(x - x^0) + b + \overline{(\boldsymbol{A} - A)(\boldsymbol{x} - x^0)}.$$

For $A := A^c$ and $x := x^0$, we particularly get

$$h(x) \geq A(x - x^0) + b - A^{\Delta} x^{\Delta},$$
$$h(x) \leq A(x - x^0) + b + A^{\Delta} x^{\Delta}.$$

Theorem 2. *For any selection of $x^0 \in \boldsymbol{x}$ and $A \in \boldsymbol{A}$, the interval linear programming approach from Section 2 yields always as tight enclosures as the parallel linearization.*

Proof. We consider the estimation from above; the estimation from below can be done accordingly. By the procedure from Section 2.2, the function $h(x)$ on $\boldsymbol{x}$ is estimated from above by

$$h(x) \leq A^c(x - x^0) + A^\Delta |x - x^0| + b.$$

(This includes the vertex selection of x^0, too.) Then, the absolute value $|x - x^0|$ is linearized by means of Beaumont

$$|x - x^0| \leq D_\alpha(x - x^0) + \beta$$

for some $\alpha \in \mathbb{R}^n$ and $\beta \in \mathbb{R}$. We want to show that the interval linear programming upper bound

$$h(x) \leq A^c(x - x^0) + A^\Delta (D_\alpha(x - x^0) + \beta) + b$$

falls into parallel linearization estimations, that is,

$$A^c(x - x^0) + A^\Delta (D_\alpha(x - x^0) + \beta) + b \in A(x - x^0) + (\boldsymbol{A} - A)(\boldsymbol{x} - x^0) + b,$$

or, equivalently,

$$(A^c - A)(x - x^0) + A^\Delta (D_\alpha(x - x^0) + \beta) \in (\boldsymbol{A} - A)(\boldsymbol{x} - x^0).$$

The ith row of this inclusion reads

$$\sum_{j=1}^n (a_{ij}^c - a_{ij})(x_j - x_j^0) + \sum_{j=1}^n a_{ij}^\Delta (\alpha_j(x_j - x_j^0) + \beta_j) \in \sum_{j=1}^n (\boldsymbol{a}_{ij} - a_{ij})(\boldsymbol{x}_j - x_j^0).$$

We prove a stronger statement, claiming that for any i, j,

$$(a_{ij}^c - a_{ij})(x_j - x_j^0) + a_{ij}^\Delta (\alpha_j(x_j - x_j^0) + \beta_j) \in (\boldsymbol{a}_{ij} - a_{ij})(\boldsymbol{x}_j - x_j^0).$$

Substituting for α_j and β_j, the left-hand side draws

$$(a_{ij}^c - a_{ij})(x_j - x_j^0) + a_{ij}^\Delta \left(\frac{|\overline{x}_j - x_j^0| - |\underline{x}_j - x_j^0|}{2x_j^\Delta} (x_j - x_j^0) \right.$$
$$\left. + \frac{(\overline{x}_j - x_j^0)|\underline{x}_j - x_j^0| - (\underline{x}_j - x_j^0)|\overline{x}_j - x_j^0|}{2x_j^\Delta} \right) \tag{14}$$

This is a linear function in x_j, so it is sufficient to show the inclusion only for both end-points of $\boldsymbol{x}_j$. Putting $x_j := \overline{x}_j$, the function (14) simplifies to

$$(a_{ij}^c - a_{ij})(\overline{x}_j - x_j^0) + a_{ij}^\Delta \left(\frac{|\overline{x}_j - x_j^0|}{2x_j^\Delta} (\overline{x}_j - x_j^0) - \frac{(\underline{x}_j - x_j^0)|\overline{x}_j - x_j^0|}{2x_j^\Delta} \right)$$
$$= (a_{ij}^c - a_{ij})(\overline{x}_j - x_j^0) + a_{ij}^\Delta |\overline{x}_j - x_j^0|$$
$$\in (\boldsymbol{a}_{ij} - a_{ij})(\overline{x}_j - x_j^0)$$
$$\subseteq (\boldsymbol{a}_{ij} - a_{ij})(\boldsymbol{x}_j - x_j^0).$$

For $x_j := \underline{x}_j$, the proof is analogous. $\qquad \square$

4   Conclusion

We showed that relaxation in constraint programming can be handled by means of interval linear programming. This approach is easily generalized for global optimization problems, too. Polyhedral relaxations are particularly convenient for problems with continuous solution sets and for high-dimensional problems. Curse of dimensionality still remains true, however, linear programming works efficiently and the polyhedral relaxation is cheap to calculate.

Our approach has some degrees of freedom concerning the choices of x^0. We recommend to choose the center and two opposite vertices of the initial box, but other choices may be just as good. If we have some information from the previous iterations, then other suitable choices of x^0 are under consideration. Basically, more choices is better since it only increases correspondingly the number of inequalities in the linear program.

Acknowledgments. The authors were supported by the Czech Science Foundation Grant P402-13-10660S, and J. Horáček in addition by the Charles University grant GAUK No. 712912.

References

1. Althaus, E., Becker, B., Dumitriu, D., Kupferschmid, S.: Integration of an LP solver into interval constraint propagation. In: Wang, W., Zhu, X., Du, D.-Z. (eds.) COCOA 2011. LNCS, vol. 6831, pp. 343–356. Springer, Heidelberg (2011)
2. Araya, I., Trombettoni, G., Neveu, B.: A contractor based on convex interval taylor. In: Beldiceanu, N., Jussien, N., Pinson, É. (eds.) CPAIOR 2012. LNCS, vol. 7298, pp. 1–16. Springer, Heidelberg (2012)
3. Baharev, A., Achterberg, T., Rév, E.: Computation of an extractive distillation column with affine arithmetic. AIChE J. 55(7), 1695–1704 (2009)
4. Beaumont, O.: Solving interval linear systems with linear programming techniques. Linear Algebra Appl. 281(1-3), 293–309 (1998)
5. Chen, L., Miné, A., Wang, J., Cousot, P.: Interval polyhedra: An abstract domain to infer interval linear relationships. In: Palsberg, J., Su, Z. (eds.) SAS 2009. LNCS, vol. 5673, pp. 309–325. Springer, Heidelberg (2009)
6. Collavizza, H., Delobel, F., Rueher, M.: Comparing partial consistencies. Reliab. Comput. 5(3), 213–228 (1999)
7. Fiedler, M., Nedoma, J., Ramík, J., Rohn, J., Zimmermann, K.: Linear optimization problems with inexact data. Springer, New York (2006)
8. Gerlach, W.: Zur Lösung linearer Ungleichungssysteme bei Störung der rechten Seite und der Koeffizientenmatrix. Math. Operationsforsch. Stat. Ser. Optimization 12, 41–43 (1981)
9. Goualard, F., Jermann, C.: A reinforcement learning approach to interval constraint propagation. Constraints 13(1), 206–226 (2008)
10. Granvilliers, L.: On the combination of interval constraint solvers. Reliab. Comput. 7(6), 467–483 (2001)
11. Hansen, E.R., Walster, G.W.: Global optimization using interval analysis, 2nd edn. Marcel Dekker, New York (2004)

12. Hladík, M.: Interval linear programming: A survey. In: Mann, Z.A. (ed.) Linear Programming - New Frontiers in Theory and Applications, ch. 2, pp. 85–120. Nova Science Publishers, New York (2012)

13. Hladík, M.: Weak and strong solvability of interval linear systems of equations and inequalities. Linear Algebra Appl. 438(11), 4156–4165 (2013)

14. Jansson, C.: Rigorous lower and upper bounds in linear programming. SIAM J. Optim. 14(3), 914–935 (2004)

15. Jaulin, L.: Reliable minimax parameter estimation. Reliab. Comput. 7(3), 231–246 (2001)

16. Jaulin, L., Kieffer, M., Didrit, O., Walter, É.: Applied interval analysis. With examples in parameter and state estimation, robust control and robotics. Springer, London (2001)

17. Kearfott, R.B.: Discussion and empirical comparisons of linear relaxations and alternate techniques in validated deterministic global optimization. Optim. Methods Softw. 21(5), 715–731 (2006)

18. Lebbah, Y., Lhomme, O.: Accelerating filtering techniques for numeric CSPs. Artif. Intell. 139(1), 109–132 (2002)

19. Lebbah, Y., Michel, C., Rueher, M.: An efficient and safe framework for solving optimization problems. J. Comput. Appl. Math. 199(2), 372–377 (2007)

20. Lebbah, Y., Michel, C., Rueher, M., Daney, D., Merlet, J.-P.: Efficient and safe global constraints for handling numerical constraint systems. SIAM J. Numer. Anal. 42(5), 2076–2097 (2005)

21. Lin, Y., Stadtherr, M.A.: LP strategy for the interval-Newton method in deterministic global optimization. Ind. Eng. Chem. Res. 43(14), 3741–3749 (2004)

22. Neumaier, A.: Interval methods for systems of equations. Cambridge University Press, Cambridge (1990)

23. Neumaier, A., Shcherbina, O.: Safe bounds in linear and mixed-integer linear programming. Math. Program. 99(2), 283–296 (2004)

24. Oettli, W., Prager, W.: Compatibility of approximate solution of linear equations with given error bounds for coefficients and right-hand sides. Numer. Math. 6, 405–409 (1964)

25. Ratschan, S., She, Z.: Providing a basin of attraction to a target region of polynomial systems by computation of Lyapunov-like functions. SIAM J. Control Optim. 48(7), 4377–4394 (2010)

26. Trombettoni, G., Araya, I., Neveu, B., Chabert, G.: Inner regions and interval linearizations for global optimization. In: Burgard, W., Roth, D. (eds.) Proceedings of the Twenty-Fifth AAAI Conference on Artificial Intelligence, AAAI 2011, San Francisco, California, USA. AAAI Press (2011)

27. Vu, X.-H., Sam-Haroud, D., Faltings, B.: Enhancing numerical constraint propagation using multiple inclusion representations. Ann. Math. Artif. Intell. 55(3-4), 295–354 (2009)

Selecting the Best Location for a Meteorological Tower: A Case Study of Multi-objective Constraint Optimization

Aline Jaimes, Craig Tweedy, Tanja Magoc,
Vladik Kreinovich, and Martine Ceberio

University of Texas at El Paso,
500 W. University,
El Paso, TX 79968, USA
vladik@utep.edu

Abstract. Using the problem of selecting the best location for a meteorological tower as an example, we show that in multi-objective optimization under constraints, the traditional weighted average approach is often inadequate. We also show that natural invariance requirements lead to a more adequate approach – a generalization of Nash's bargaining solution.

Case Study. We want to select the best location of a sophisticated multi-sensor meteorological tower. We have several criteria to satisfy.

For example, the station should not be located too close to a road, so that the gas flux generated by the cars do not influence our measurements of atmospheric fluxes; in other words, the distance x_1 to the road should be larger than a certain threshold t_1: $x_1 > t_1$, or $y_1 \overset{\text{def}}{=} x_1 - t_1 > 0$.

Also, the inclination x_2 at the should be smaller than a corresponding threshold t_2, because otherwise, the flux will be mostly determined by this inclination and will not be reflective of the atmospheric processes: $x_2 < t_2$, or $y_2 \overset{\text{def}}{=} t_2 - x_2 > 0$.

General Case. In general, we have several such differences $y_1, \ldots, y_n$ all of which have to be non-negative. For each of the differences y_i, the larger its value, the better.

Multi-criteria Optimization. Our problem is a typical setting for *multi-criteria optimization*; see, e.g., [1, 4, 5].

Weighted Average. A most widely used approach to multi-criteria optimization is *weighted average*, where we assign weights $w_1, \ldots, w_n > 0$ to different criteria y_i and select an alternative for which the weighted average $w_1 \cdot y_1 + \ldots + w_n \cdot y_n$ attains the largest possible value.

M. Ceberio and V. Kreinovich (eds.), *Constraint Programming and Decision Making*,   61
Studies in Computational Intelligence 539,
DOI: 10.1007/978-3-319-04280-0_7, © Springer International Publishing Switzerland 2014

Additional Requirement. In our problem, we have an additional requirement – that all the values y_i must be positive. Thus, we must only compare solutions with $y_i > 0$ when selecting an alternative with the largest possible value of the weighted average.

Limitations of the Weighted Average Approach. In general, the weighted average approach often leads to reasonable solutions of the multi-criteria optimization problem. However, as we will show, in the presence of the additional positivity requirement, the weighted average approach is not fully satisfactory.

A Practical Multi-criteria Optimization Must Take into Account That Measurements Are Not Absolutely Accurate. Indeed, the values y_i come from measurements, and measurements are never absolutely accurate. The results $\widetilde{y}_i$ of the measurements are close to the actual (unknown) values y_i of the measured quantities, but they are not exactly equal to these values. If

- we measure the values y_i with higher and higher accuracy and,
- based on the resulting measurement results $\widetilde{y}_i$, we conclude that the alternative $y = (y_1, \ldots, y_n)$ is better than some other alternative $y' = (y'_1, \ldots, y'_n)$,

then we expect that the actual alternative y is indeed either better than y' or at least of the same quality as y'. Otherwise, if we do not make this assumption, we will not be able to make any meaningful conclusions based on real-life (approximate) measurements.

The Above Natural Requirement Is Not Always Satisfied for Weighted Average. Let us show that for the weighted average, this "continuity" requirement is not satisfied even in the simplest case when we have only two criteria y_1 and y_2. Indeed, let $w_1 > 0$ and $w_2 > 0$ be the weights corresponding to these two criteria. Then, the resulting strict preference relation $\succ$ has the following properties:

- if $y_1 > 0$, $y_2 > 0$, $y'_1 > 0$, and $y'_2 > 0$, and $w_1 \cdot y_1 + w_2 \cdot y_2 > w_1 \cdot y'_1 + w_2 \cdot y'_2$, then

$$y = (y_1, y_2) \succ y' = (y'_1, y'_2);$$

- if $y_1 > 0$, $y_2 > 0$, and at least one of the values y'_1 and y'_2 is non-positive, then

$$y = (y_1, y_2) \succ y' = (y'_1, y'_2).$$

Let us consider, for every $\varepsilon > 0$, the tuple $y(\varepsilon) \stackrel{\text{def}}{=} \left(\varepsilon, 1 + \dfrac{w_1}{w_2} \right)$, with $y_1(\varepsilon) = \varepsilon$ and $y_2(\varepsilon) = 1 + \dfrac{w_1}{w_2}$, and also the comparison tuple $y' = (1, 1)$. In this case, for every $\varepsilon > 0$, we have

$$w_1 \cdot y_1(\varepsilon) + w_2 \cdot y_2(\varepsilon) = w_1 \cdot \varepsilon + w_2 + w_2 \cdot \frac{w_1}{w_2} = w_1 \cdot (1 + \varepsilon) + w_2$$

and

$$w_1 \cdot y'_1 + w_2 \cdot y'_2 = w_1 + w_2,$$

hence $y(\varepsilon) \succ y'$. However, in the limit $\varepsilon \to 0$, we have $y(0) = \left(0, 1 + \dfrac{w_1}{w_2}\right)$, with $y(0)_1 = 0$ and thus, $y(0) \prec y'$.

What We Want: A Precise Description. We want to be able to compare different alternatives.

Each alternative is characterized by a tuple of n values $y = (y_1, \ldots, y_n)$, and only alternatives for which all the values y_i are positive are allowed. Thus, from the mathematical viewpoint, the set of all alternatives is the set $(R^+)^n$ of all the tuples of positive numbers.

For each two alternatives y and y', we want to tell whether y is better than y' (we will denote it by $y \succ y'$ or $y' \prec y$), or y' is better than y ($y' \succ y$), or y and y' are equally good ($y' \sim y$). These relations must satisfy natural properties. For example, if y is better than y' and y' is better than y'', then y is better than y''. In other words, the relation $\succ$ must be transitive. Similarly, the relation $\sim$ must be transitive, symmetric, and reflexive ($y \sim y$), i.e., in mathematical terms, an *equivalence relation*.

So, we want to define a pair of relations $\succ$ and $\sim$ such that $\succ$ is transitive, $\sim$ is transitive, $\sim$ is an equivalence relation, and for every y and y', one and only one of the following relations hold: $y \succ y'$, $y' \succ y$, or $y \sim y'$.

It is also reasonable to require that if each criterion is better, then the alternative is better as well, i.e., that if $y_i > y'_i$ for all i, then $y \succ y'$.

Comment. Pairs of relations of the above type can be alternatively characterized by a *pre-ordering* relation

$$a \succeq b \Leftrightarrow (a \succ b \vee a \sim b).$$

This relation must be transitive and – in our case – total (i.e., for every y and y', we have $y \succeq y' \vee y' \succeq y$. Once we know the pre-ordering relation $\succeq$, we can reconstruct $\succ$ and $\sim$ as follows:

$$y \succ y' \Leftrightarrow (y \succeq y' \,\&\, y' \not\succeq y);$$

$$y \sim y' \Leftrightarrow (y \succeq y' \,\&\, y' \succeq y).$$

Scale Invariance: Motivation. The quantities y_i describe completely different physical notions, measured in completely different units. In our meteorological case, some of these values are wind velocities measured in meters per second, or in kilometers per hour, or miles per hour. Other values are elevations described in meters, kilometers, or feet, etc. Each of these quantities can be described in many different units. A priori, we do not know which units match each other, so it is reasonable to assume that the units used for measuring different quantities may not be exactly matched.

It is therefore reasonable to require that the relations $\succ$ and $\sim$ between the two alternatives $y = (y_1, \ldots, y_n)$ and $y' = (y'_1, \ldots, y'_n)$ do not change if we simply change the units in which we measure each of the corresponding n quantities.

Scale Invariance: Towards a Precise Description. When we replace a unit in which we measure a certain quantity q by a new measuring unit which is $\lambda > 0$ times smaller, then the numerical values of this quantity increase by a factor of λ: $q \to \lambda \cdot q$. For example, 1 cm is $\lambda = 100$ times smaller than 1 m, so the length $q = 2$ m, when measured in cm, becomes $\lambda \cdot q = 2 \cdot 100 = 200$ cm.

Let λ_i denote the ratio of the old to the new units corresponding to the i-th quantity. Then, the quantity that had the value y_i in the old units will be described by a numerical value $\lambda_i \cdot y_i$ in the new unit. Therefore, scale-invariance means that for all $y, y \in (R^+)^n$ and for all $\lambda_i > 0$, we have

$$y = (y_1, \ldots, y_n) \succ y' = (y'_1, \ldots, y'_n) \Rightarrow (\lambda_1 \cdot y_1, \ldots, \lambda_n \cdot y_n) \succ (\lambda_1 \cdot y'_1, \ldots, \lambda_n \cdot y'_n)$$

and

$$y = (y_1, \ldots, y_n) \sim y' = (y'_1, \ldots, y'_n) \Rightarrow (\lambda_1 \cdot y_1, \ldots, \lambda_n \cdot y_n) \sim (\lambda_1 \cdot y'_1, \ldots, \lambda_n \cdot y'_n).$$

Continuity. As we have mentioned in the previous section, we also want to require that the relations $\succ$ and $\sim$ are *continuous* in the following sense: if $y(\varepsilon) \succeq y'(\varepsilon)$ for every ε, then in the limit, when $y(\varepsilon) \to y(0)$ and $y'(\varepsilon) \to y'(0)$ (in the sense of normal convergence in R^n), we should have $y(0) \succeq y'(0)$.

Let us now describe our requirements in precise terms.

Definition 1. *By a* total pre-ordering relation *on a set Y, we mean a pair of a transitive relation $\succ$ and an equivalence relation $\sim$ for which, for every $y, y' \in Y$, one and only one of the following relations hold: $y \succ y'$, $y' \succ y$, or $y \sim y'$.*

Comment. We will denote $y \succeq y' \overset{\text{def}}{=} (y \succ y' \vee y \sim y')$.

Definition 2. *We say that a total pre-ordering is* non-trivial *if there exist y and y' for which $y \succ y'$.*

Comment. This definition excludes the trivial pre-ordering in which every two tuples are equivalent to each other.

Definition 3. *We say that a total pre-ordering relation on the set $(R^+)^n$ is:*

- monotonic *if $y'_i > y_i$ for all i implies $y' \succ y$;*
- scale-invariant *if for all $\lambda_i > 0$:*
  - $(y'_1, \ldots, y'_n) \succ y = (y_1, \ldots, y_n)$ *implies* $(\lambda_1 \cdot y'_1, \ldots, \lambda_n \cdot y'_n) \succ (\lambda_1 \cdot y_1, \ldots, \lambda_n \cdot y_n)$, *and*
  - $(y'_1, \ldots, y'_n) \sim y = (y_1, \ldots, y_n)$ *implies* $(\lambda_1 \cdot y'_1, \ldots, \lambda_n \cdot y'_n) \sim (\lambda_1 \cdot y_1, \ldots, \lambda_n \cdot y_n)$.
- continuous *if whenever we have a sequence $y^{(k)}$ of tuples for which $y^{(k)} \succeq y'$ for some tuple y', and the sequence $y^{(k)}$ tends to a limit y, then $y \succeq y'$.*

Theorem. *Every non-trivial monotonic scale-invariant continuous total pre-ordering relation on* $(R^+)^n$ *has the following form:*

$$y' = (y'_1, \ldots, y'_n) \succ y = (y_1, \ldots, y_n) \Leftrightarrow \prod_{i=1}^{n} (y'_i)^{\alpha_i} > \prod_{i=1}^{n} y_i^{\alpha_i};$$

$$y' = (y'_1, \ldots, y'_n) \sim y = (y_1, \ldots, y_n) \Leftrightarrow \prod_{i=1}^{n} (y'_i)^{\alpha_i} = \prod_{i=1}^{n} y_i^{\alpha_i},$$

for some constants $\alpha_i > 0$.

Comment. In other words, for every non-trivial monotonic scale-invariant continuous total pre-ordering relation on $(R^+)^n$, there exist values $\alpha_1 > 0$, ..., $\alpha_n > 0$ for which the above equivalence hold. Vice versa, for each set of values $\alpha_1 > 0$, ..., $\alpha_n > 0$, the above formulas define a monotonic scale-invariant continuous pre-ordering relation on $(R^+)^n$.

It is worth mentioning that the resulting relation coincides with the asymmetric version [3] of the bargaining solution proposed by the Nobelist John Nash in 1953 [2].

Acknowledgments. This work was supported in part by the National Science Foundation grant HRD-0734825.

References

1. Ehrgott, M., Gandibleux, X. (eds.): Multiple Criteria Optimization: State of the Art Annotated Bibliographic Surveys. Springer, Heidelberg (2002)
2. Nash, J.: Two-Person Cooperative Games. Econometrica 21, 128–140 (1953)
3. Roth, A.: Axiomatic Models of Bargaining. Springer, Berlin (1979)
4. Sawaragi, Y., Nakayama, H., Tanino, T.: Theory of Multiobjective Optimization. Academic Press, Orlando (1985)
5. Steuer, E.E.: Multiple Criteria Optimization: Theory, Computations, and Application. John Wiley & Sons, New York (1986)

Gibbs Sampling as a Natural Statistical Analog of Constraints Techniques: Prediction in Science under General Probabilistic Uncertainty

Misha Koshelev

Human Neuroimaging Lab,
Baylor College of Medicine,
One Baylor Plaza S104,
Houston, TX 77030, USA
misha.koshelev@bcm.edu

Abstract. One of the main objectives of science is to predict future events, i.e., more precisely, the results of future measurements and observations. If we take into account the probabilistic uncertainty related to the inaccuracy of the measurement results, to the inaccuracy of the model, and to the inaccuracy of the prior information, then the most adequate approach is to generate a posterior distribution by using Bayes' theorem. For the simplest posterior distributions, we can deduce explicit analytical formulas for the resulting statistical characteristics (mean, standard deviation, etc.) of the predict future measurement result. However, in general, such formulas are not possible, so we have to use a Monte-Carlo simulation of the corresponding joint distribution of the future measurement results and model parameters.

The main computational challenge here is that there is no general algorithm for simulating an arbitrary multi-variate distribution; such algorithms are known only for single-variate distributions and – in some cases – for the case of several variables. Thus, we need to reduce the general simulation problem to such simplified cases. We show that this problem can be solved by using the general constraints approach, and that this idea clarifies Gibbs sampling – one of the most widely used techniques for such simulation. This interpretation of Gibbs sampling enables us to analyze Gibbs sampling – in particular, to obtain a (somewhat counterintuitive) result that while a straightforward parallelization is possible for deterministic constraint propagation, parallelization does not work even in the simplest two-variable probabilistic case.

Formulation of the Problem: Prediction in Science in a Realistic Setting – under Probabilistic Uncertainty. One of the main objectives of science is to predict future events – and thus, if we have a choice, to come up with a choice which leads to the most beneficial future situation.

To predict an event means to predict the values of different observable and measurable quantities q. In order to predict these values, we must know how these

M. Ceberio and V. Kreinovich (eds.), *Constraint Programming and Decision Making*,   67
Studies in Computational Intelligence 539,
DOI: 10.1007/978-3-319-04280-0_8, © Springer International Publishing Switzerland 2014

values depend on time t. For this dependence, we usually have a *model* $q(t) = f(t, c_1, \ldots, c_n)$, where f is an algorithmically given function, and $c_1, \ldots, c_n$ are parameters that needs to be determined based on the previous observations and measurement results.

For example, in Newton's celestial mechanics, the parameters c_i are the current masses, coordinates, and velocities of the celestial bodies, and the algorithm f for computing the i-th coordinate $x_{ai}(t)$ of the a-th body at moment t consists of integrating the corresponding differential equations of motion

$$m_a \cdot \frac{d^2 x_{ai}}{dt^2} = \vec{F}_a = \sum_{b \neq a} \frac{m_b \cdot m_a}{|\vec{x}_b - \vec{x}_a|^3} \cdot (x_{bi} - x_{ai}).$$

In Newton's theory, the formulas predict the exact values of the coordinates. However, what we really want to predict are not the actual (unknown) future values of the coordinates (or other physical quantities) but rather the (potentially observable) future results of measuring these quantities. Since measurements are never absolutely accurate, the measurement result is usually somewhat different from the actual value of the measured quantity. At best, even if we know the exact actual values of the future quantities, we can predict the *probabilities* of different measurement results. Thus, at best, the model enables us, given the values of the parameters $c_1, \ldots, c_n$, to predict the *probability* of different measurement results.

This is especially true for situations like statistical physics where even the exact future values of different quantities cannot be predicted: we can only only predict the probabilities of different future values.

In general, the model enables us, given parameters values $\vec{c}$, to predict the probabilities of different measurement results.

Traditional Physical Approach and Its Limitations. In the traditional physical approach, we first estimate the values of the parameters c_i based on the results of the previous observations and measurements, and then we use these estimated values to compute the probabilities of different future results.

The main limitation of this approach is that it does not take into account the fact that the estimates c_i are approximations. Let us illustrate this limitation on the simplified example when we have exactly one parameter c_1, and the actual value coincides with this parameter: $q_1 = c_1$. In the past, all the measurements were with a significant measurement inaccuracy. As a result, from these measurement, we only get an approximate value $\widetilde{c}_1$ of the parameter c_1. Let us assume that the future measurement is, in contrast, very accurate, so its measurement inaccuracy can be safely ignored.

In this case, the traditional physical approach predicts that the measured value will be exactly $\widetilde{c}_1$. Thus, the predicted probability of getting this value is 1, and the predicted probability of getting any other value $c_1 \neq \widetilde{c}_1$ is 0. In reality, of course, the measured value will be equal to a slightly different number c_1. Thus, in reality, we will observe the measurement value whose predicted probability is 0.

In general, it is therefore desirable to take into account the difference between the estimates $\tilde{c}_i$ and the actual values c_i when predicting the probabilities of different future measurement results.

Another limitation is that the traditional physical approach does not take into account that, in addition to observations and measurements, we often have additional *prior* information about the probability of different values $\vec{c}$.

Statistical Way to Take Prior Information into Account: Bayesian Approach. In decision theory, it has been shown that under reasonable assumptions [3, 4], each prior information can be formulated as an appropriate "prior" probability distribution $\rho_0(\vec{c})$ on the set of all possible values of the parameters $\vec{c} = (c_1, \ldots, c_n)$. In addition to this prior probability, we have a *model*, that, for any given $\vec{c}$, predicts the probability (density) of different measured values E: $\rho(E \mid \vec{c})$. In this situation, once we know all the values E of the previous observations and measurements, we can estimate the resulting probability $\vec{c}$ by using the Bayes' formula

$$\rho(\vec{c} \mid E) = \frac{1}{N} \cdot \rho(E \mid \vec{c}) \cdot \rho_0(\vec{c}),$$

where N is the normalization coefficient selected to guarantee that the overall probability is 1: $\int \rho(\vec{c}) \, d\vec{c} = 1$, i.e.,

$$N = \int \rho(E \mid \vec{c}) \cdot \rho_0(\vec{c}) \, d\vec{c}.$$

Now, the probability of different values $q(t)$ can be obtained from the formula of full probability, by combining the probability of getting this value q for different parameters $\vec{c}$:

$$\rho_q(q) = \int \rho(q \mid \vec{c}) \cdot \rho(\vec{c} \mid E) \, d\vec{c}.$$

Bayesian Approach: Need for Monte-Carlo Simulations. Our objective is to estimate such characteristics as the expected value of the predicted quantity, the standard deviation, etc.

In the simplest cases, e.g., when all the distributions are Gaussian and independent, it is usually possible to come up with explicit analytical formulas for these characteristics. However, in the general case, it is not possible to have analytical formulas. In this case, under probabilistic uncertainty, it is reasonable to use Monte-Carlo approach, in which we simulate the distribution of all the involved unknown quantities: the parameters c_i, and the future measured values q, according to the joint distribution

$$\rho(\vec{c}, q) = \frac{1}{N} \cdot \rho(q \mid \vec{c}) \cdot \rho(E \mid \vec{c}) \cdot \rho_0(\vec{c}).$$

Once we have simulated this distribution, i.e., once we have a sample $(\vec{c}^{(k)}, q^{(k)})$, $(k = 1, \ldots, M)$ whose distribution follows the above probability distribution law, we can use the values $q^{(k)}$ from the corresponding simulated sample to estimate the mean $E[q]$, the standard deviation $\sigma[q]$, and all other characteristics of q by using the usual formulas

$$E[q] \approx \frac{1}{M} \cdot \sum_{k=1}^{M} q^{(k)}, \quad (\sigma[q])^2 = \frac{1}{M} \cdot \sum_{k=1}^{M} (q^{(k)} - E[q])^2.$$

Monte-Carlo Simulation: Computational Challenge. The main computational challenge is that there is no general way, even when we have an analytical formula for the joint distribution, to produce the sample of tuples distributed according to this distribution.

Monte-Carlo Simulation: Cases for Which Algorithms Are Known. While there is no general algorithm for simulating an arbitrary probability distribution, it is algorithmically possible to simulate an arbitrary distribution of a *single* random variable. It is also sometimes algorithmically possible to simulate a joint distribution for several variables: e.g., when this joint distribution is Gaussian.

In a computer, there is usually a standard number generator that generates numbers r uniformly distributed on the interval $[0, 1]$. Thus, we can simulate an arbitrary distribution by reducing it to this standard one. These are two known algorithms for this reduction. In the first algorithm, we assume that we know the cumulative distribution function (cdf) $F(X) = \text{Prob}(x \leq X)$ – and that we know the corresponding inverse function $F^{-1}(u)$ for which $F(F^{-1}(u)) = u$. In this case, the result $F^{-1}(r)$ of applying this inverse function to the result r of the standard random generator is distributed according to the desired distribution $F(X)$.

In the second algorithm, we assume that the distribution is located on an interval $[\underline{X}, \overline{X}]$, and that we know the probability density function (pdf) $\rho(x)$, and we know its largest value ρ_0 on this interval. Under this assumption, we can simulate the random variable uniformly distributed on the interval $[\underline{X}, \overline{X}]$ as $x = \underline{X} + r \cdot (\overline{X} - \underline{X})$, then simulate again the standard random number generator r, and pick x if $r \leq \rho(x)/\rho_0$. One can see that in this case, the probability of selecting each value $x \in [\underline{X}, \overline{X}]$ is indeed proportional to the desired pdf $\rho(x)$.

Constraint Propagation: Brief Reminder. Let us show that to solve the above problem, we can use the ideas from constraint propagation. Indeed, in constraint propagation, we are interested in finding the values of the *deterministic* variables $x_1, \ldots, x_n$ that satisfies the given constraints, e.g., constraints of the type $f_i(x_1, \ldots, x_n) = 0$ or $f_j(x_1, \ldots, x_n) \geq 0$.

In the constraint propagation algorithms, we first transform each constraint into an equivalent sequence of simpler constraints, i.e., constraints which are simple enough so that for each of the resulting constraints $g_i(x_1, \ldots, x_n) = 0$ and for each variable j, once we know the values of all other variables $x_1, \ldots, x_{j-1}, x_{j+1}, \ldots, x_n$, we can algorithmically find the value x_j for which this constraint is satisfied (or, alternatively, the set – usually, an interval – of possible values x_j for which this constraint is satisfied).

Also, once we know the intervals (or more general sets) $\mathbf{x}_1, \ldots, \mathbf{x}_{j-1}, \mathbf{x}_{j+1}, \ldots,$ $\mathbf{x}_n$ of possible values of the corresponding variables $x_1, \ldots, x_{j-1}, x_{j+1}, \ldots, x_n,$ we can estimate the interval (set) $\mathbf{x}_j$ of possible related values x_j – possible in the sense that the desired constraint is satisfies for some $x_1 \in \mathbf{x}_1, \ldots, x_{j-1} \in \mathbf{x}_{j-1},$ $x_{j+1} \in \mathbf{x}_{j+1}, \ldots, x_n \in \mathbf{x}_n.$

Once we have a list of such simplified constraints, we repeatedly use the corresponding value-determining algorithm to find the value (or set of values) of different variables based on what we have already computed for the others. In the numerical (no-sets) version of this algorithm, at each iteration t, we select a constraint $g_i(x_1, \ldots, x_n) = 0$ and a variable x_j, and we use the above idea to find the new estimate $x_j^{[t]}$ for x_j based on the previous estimates $x_1^{[t-1]}, \ldots, x_{j-1}^{[t-1]}, x_{j+1}^{[t-1]}, \ldots, x_n^{[t-1]}$ of all the other variables. In other words, we find the value $x_j^{[t]}$ for which

$$g_i(x_1^{[t-1]}, \ldots, x_{j-1}^{[t-1]}, x_j^{[t]}, x_{j+1}^{[t-1]}, \ldots, x_n^{[t-1]}) = 0.$$

For all other variables x_k, $k \neq j$, we keep the previous values: $x_k^{[t]} = x_k^{[t-1]}$.

If the Process Converges, It Converges to the Desired Values. If this process converges (i.e., if $x_k^{[t]} \to x_k$ for all k), then in the limit, we conclude that $g_i(x_1, \ldots, x_n) = 0$, i.e., that the i-th constraint is satisfied. Since we are constantly cycling through all the constraints, this means that in the limit, we satisfy all the constraints, so the limit tuple indeed solves the original constraint satisfaction problem.

Similarly, if we deal with sets of possible values and each of these sets tends to a single value, then these limit values satisfy all the desired constraints; see, e.g., [2].

A Natural Probabilistic Analogue of Constraint Propagation. In constraint propagation, once we know the values of all the variables $x_1, \ldots, x_{j-1}, x_{j+1}, \ldots, x_n$ except for one x_j, we can then determine either the value x_j of the selected variable – or, if we cannot determine x_j uniquely, we can find the *set* of possible values of x_j.

In the probabilistic case, once we know the values $x_1, \ldots, x_{j-1}, x_{j+1}, \ldots, x_n$ of all the variables except for the selected one, we cannot determine the remaining value x_j uniquely. Instead, we can find the corresponding conditional probability distribution for this remaining variable, with the conditional density $\rho(x_j \mid x_1, \ldots, x_{j-1}, x_{j+1}, \ldots, x_n)$. Since this resulting distribution is a single-variate distribution, we can use one of the techniques for simulating this distribution and get the corresponding x_j. Thus, we arrive at the following algorithm for simulating an arbitrary multi-variate distribution.

To generate one tuple $(x_1, \ldots, x_n)$ from the desired sample, we start with an arbitrary tuple $x_1^{[0]}, \ldots, x_n^{[0]}$. On each iteration t, we select a variable x_j, and use the 1-D Monte-Carlo simulation to generate a value

$x_j^{[t]}$ distributed according to the corresponding conditional distribution $\rho(x_j \,|\, x_1^{[t-1]}, \ldots, x_{j-1}^{[t-1]}, x_{j+1}^{[t-1]}, \ldots, x_n^{[t-1]})$.

For all other variables x_k, $k \neq j$, we keep the previous values: $x_k^{[t]} = x_k^{[t-1]}$. We make sure that each variable is periodically selected: e.g., by simply cycling through the variables in their natural order: first, we select x_1, then x_2, $\ldots$, then x_n, then x_1 again, etc.

Comment. This iterative process is known and it is one of the most widely used in Monte-Carlo simulations, especially in the statistical analysis of human behavior experiments (see, e.g., [1] and references therein). It is called *Gibbs sampling* because it was originally derived from a set of complex ideas related to Gibbs distribution in statistical physics. We have shown that it can be easier (and, we believe, more naturally) derived if we view the simulation problem as a natural probabilistic analogue of the constraint problems.

If the Process Converges, It Converges to the Desired Distribution. Let us show that, similarly to the usual case of deterministic constraints, in the probabilistic case, if the process converges, i.e., if the probability distribution of the tuples on each iteration converges to some limit distribution $\rho_l(x_1, \ldots, x_n)$, then this limit distribution coincides with the original distribution $\rho(x_1, \ldots, x_n)$.

Indeed, in the limit, since the limit distribution of equal to $\rho_l(x_1, \ldots, x_n)$, the conditional distribution of x_j relative to all the other variables has the corresponding conditional probability density

$$\rho_l(x_j \,|\, x_1, \ldots, x_{j-1}, x_{j+1}, \ldots, x_n) = \frac{\rho_l(x_1, \ldots, x_{j-1}, x_j, x_{j+1}, \ldots, x_n)}{m_l(x_1, \ldots, x_{j-1}, x_{j+1}, \ldots, x_n)},$$

where

$$m_l(x_1, \ldots, x_{j-1}, x_{j+1}, \ldots, x_n) \stackrel{\text{def}}{=} \int \rho_l(x_1, \ldots, x_{j-1}, x_j, x_{j+1}, \ldots, x_n) \, dx_j$$

is the corresponding marginal distribution.

On the other hand, according to our iterative process, in the limit, the probability of having x_j based on given values $x_1, \ldots, x_{j-1}, x_{j+1}, \ldots, x_n$ is given by the conditional probability distribution

$$\rho(x_j \,|\, x_1, \ldots, x_{j-1}, x_{j+1}, \ldots, x_n) = \frac{\rho(x_1, \ldots, x_{j-1}, x_j, x_{j+1}, \ldots, x_n)}{m(x_1, \ldots, x_{j-1}, x_{j+1}, \ldots, x_n)},$$

where $m(x_1, \ldots, x_{j-1}, x_{j+1}, \ldots, x_n)$ is the corresponding marginal distribution.

Thus, for all possible values $x_1, \ldots, x_{j-1}, x_j, x_{j+1}, \ldots, x_n$ and for all possible values j, we have

$$\rho_l(x_j \,|\, x_1, \ldots, x_{j-1}, x_{j+1}, \ldots, x_n) = \rho(x_j \,|\, x_1, \ldots, x_{j-1}, x_{j+1}, \ldots, x_n),$$

i.e.,

$$\frac{\rho_l(x_1, \ldots, x_{j-1}, x_j, x_{j+1}, \ldots, x_n)}{m_l(x_1, \ldots, x_{j-1}, x_{j+1}, \ldots, x_n)} = \frac{\rho(x_1, \ldots, x_{j-1}, x_j, x_{j+1}, \ldots, x_n)}{m(x_1, \ldots, x_{j-1}, x_{j+1}, \ldots, x_n)}.$$

From this equality, we can conclude that

$$\frac{\rho_l(x_1,\ldots,x_{j-1},x_j,x_{j+1},\ldots,x_n)}{\rho(x_1,\ldots,x_{j-1},x_j,x_{j+1},\ldots,x_n)} = \frac{m_l(x_1,\ldots,x_{j-1},x_{j+1},\ldots,x_n)}{m(x_1,\ldots,x_{j-1},x_{j+1},\ldots,x_n)},$$

i.e., that the ratio $\rho_l(\vec{x})/\rho(\vec{x})$ does not depend on x_j. Since this is true for all j, this means that this ratio does not depend on anything, i.e., it is a constant: $\rho_l(x_1,\ldots,x_{j-1},x_j,x_{j+1},\ldots,x_n) = c \cdot \rho(x_1,\ldots,x_{j-1},x_j,x_{j+1},\ldots,x_n)$. Since for both distribution, the total probability is 1, we get

$$1 = \int \rho_l(x_1,\ldots,x_{j-1},x_j,x_{j+1},\ldots,x_n)\, dx_1\ldots dx_n =$$

$$c \cdot \int \rho(x_1,\ldots,x_{j-1},x_j,x_{j+1},\ldots,x_n)\, dx_1\ldots dx_n = c \cdot 1 = c.$$

So, $c = 1$ and

$$\rho_l(x_1,\ldots,x_{j-1},x_j,x_{j+1},\ldots,x_n) = \rho(x_1,\ldots,x_{j-1},x_j,x_{j+1},\ldots,x_n).$$

In the Deterministic Case, the Standard Constraint Propagation Algorithm Can Be Parallelized. Sometimes, constraint propagation algorithms converge slowly, so it is desirable to speed up the corresponding algorithms. A natural way to speed up an algorithm is to parallelize it, i.e., to perform different computation steps in parallel. In the above algorithm, a seemingly natural way to parallelize is to change several variables in parallel: namely, once we have the values $x_1^{[t-1]}$, ..., $x_n^{[t-1]}$, we simultaneously run two or more value-determining algorithms to find the values of two or more variables $x_j^{[t]}$, $x_{j'}^{[t]}$, ... In other words, we find the values $x_j^{[t]}$, $x_{j'}^{[t]}$, ..., for which

$$g_i(x_1^{[t-1]},\ldots,x_{j-1}^{[t-1]},x_j^{[t]},x_{j+1}^{[t-1]},\ldots,x_{j'-1}^{[t-1]},x_{j'}^{[t-1]},x_{j'+1}^{[t-1]},\ldots,x_n^{[t-1]}) = 0;$$

$$g_{i'}(x_1^{[t-1]},\ldots,x_{j-1}^{[t-1]},x_j^{[t-1]},x_{j+1}^{[t-1]},\ldots,x_{j'-1}^{[t-1]},x_{j'}^{[t]},x_{j'+1}^{[t-1]},\ldots,x_n^{[t-1]}) = 0.$$

If the process converges, then in the limit, we still have

$$g_i(x_1,\ldots,x_{j-1},x_j,x_{j+1},\ldots,x_{j'-1},x_{j'},x_{j'+1},\ldots,x_n) = 0$$

and

$$g_{i'}(x_1,\ldots,x_{j-1},x_j,x_{j+1},\ldots,x_{j'-1},x_{j'},x_{j'+1},\ldots,x_n) = 0,$$

i.e., satisfaction of all the constraints.

In the Statistical Case, Parallelization Is Not Possible. Let us show that in the statistical case, in general, parallelization is not possible. Indeed, let us consider the simplest case of a 2-dimensional normal distribution. Let us assume that we have two variables x_1 and x_2 each of which is normally distributed with mean 0 and standard deviation 1, and that the covariance is equal to $\alpha \in (0,1)$.

In this case, once can show that once we know x_1, we can find x_2 as

$$x_2 = \alpha \cdot x_1 + \beta \cdot \xi,$$

where $\beta = \sqrt{1 - \alpha^2}$ and ξ is a new normally distributed random variable with mean 0 and standard deviation 1. Similarly, once we know x_2, we can find x_1 as $x_1 = \alpha \cdot x_2 + \beta \cdot \xi$.

And indeed, we can design a Gibbs sampling algorithm for simulating the corresponding distribution: we start with arbitrary values $x_1^{[0]}$ and $x_2^{[0]}$, and then alternatively replace $x_1^{[t-1]}$ and $x_2^{[t-1]}$ either with

$$x_2^{[t]} = \alpha \cdot x_1^{[t-1]} + \beta \cdot \xi^{[t]}, \quad x_1^{[t]} = x_1^{[t-1]}$$

or with

$$x_1^{[t]} = \alpha \cdot x_2^{[t-1]} + \beta \cdot \xi^{[t]}, \quad x_2^{[t]} = x_2^{[t-1]}.$$

At first, it may seem natural to parallelize this process and update both values on each step:

$$x_1^{[t]} = \alpha \cdot x_2^{[t-1]} + \beta \cdot \xi_1^{[t]}, \quad x_2^{[t]} = \alpha \cdot x_1^{[t-1]} + \beta \cdot \xi_2^{[t]}.$$

However, in this case, even if at the $(t-1)$-st step, we get the correct covariance $C^{[t-1]} = E\left[x_1^{[t-1]} \cdot x_2^{[t-1]}\right] = \alpha$, on the next step, the covariance will be

$$C^{[t]} = E\left[x_1^{[t]} \cdot x_2^{[t]}\right] = E\left[\left(\alpha \cdot x_2^{[t-1]} + \beta \cdot \xi_1^{[t]}\right) \cdot \left(\alpha \cdot x_1^{[t-1]} + \beta \cdot \xi_2^{[t]}\right)\right].$$

Since ξ_i are independent from each other and from $x_i^{[t-1]}$, we thus get

$$C^{[t]} = E\left[x_1^{[t]} \cdot x_2^{[t]}\right] = \alpha^2 \cdot E\left[x_1^{[t-1]} \cdot x_2^{[t-1]}\right] = \alpha^3 \neq \alpha.$$

Thus, even if the process converges, the limit distribution is wrong. Actually, in the above example, the covariance will decrease from α to α^3, to $(\alpha^3)^3 = \alpha^9$, ..., and, in the limit, to 0 – i.e., in the limit, instead of the correlated normal random variables, we get independent ones.

References

1. Houser, D., Keane, M., McCabe, K.: Behavior in a dynamic decision problem: An analysis of experimental evidence using a bayesian type classification algorithm. Econometrica 72(3), 781–822 (2004)
2. Jaulin, L., Kieffer, M., Didrit, O., Walter, E.: Applied Interval Analysis, with Examples in Parameter and State Estimation, Robust Control and Robotics. Springer, London (2001)
3. Jaynes, E.T.: Probability Theory: The Logic of Science, vol. 1. Cambridge University Press, Cambridge (2003)
4. Luce, R.D., Raiffa, H.: Games and Decisions: Introduction and Critical Survey. Dover, New York (1989)

Why Tensors?

Olga Kosheleva, Martine Ceberio, and Vladik Kreinovich

University of Texas at El Paso,
500 W. University,
El Paso, TX 79968, USA
{olgak,mceberio,vladik}@utep.edu

Abstract. We show that in many application areas including soft constraints reasonable requirements of scale-invariance lead to polynomial (tensor-based) formulas for combining degrees (of certainty, of preference, etc.)

Partial Orders Naturally Appear in Many Application Areas. One of the main objectives of science and engineering is to help people select decisions which are the most beneficial to them. To make these decisions,

- we must know people's preferences,
- we must have the information about different events – possible consequences of different decisions, and
- since information is never absolutely accurate and precise, we must also have information about the degree of certainty.

All these types of information naturally lead to partial orders:

- For preferences, $a < b$ means that b is preferable to a. This relation is used in *decision theory*; see, e.g., [1].
- For events, $a < b$ means that a can influence b. This causality relation is used in *space-time physics*.
- For uncertain statements, $a < b$ means that a is less certain than b. This relation is used in logics describing uncertainty such as fuzzy logic (see, e.g., [3]) and in soft constraints.

Numerical Characteristics Related to Partial Orders. While an order may be a natural way of describing a relation, orders are difficult to process, since most data processing algorithms process numbers. Because of this, in all three application areas, numerical characteristics have appeared that describe the corresponding orders:

- in decision making, *utility* describes preferences:

$$a < b \text{ if and only if } u(a) < u(b);$$

- in space-time physics, *metric* (and time coordinates) describes causality relation;
- in logic and soft constraints, numbers from the interval $[0, 1]$ are used to describe degrees of certainty; see, e.g., [3].

M. Ceberio and V. Kreinovich (eds.), *Constraint Programming and Decision Making,*
Studies in Computational Intelligence 539,
DOI: 10.1007/978-3-319-04280-0_9, © Springer International Publishing Switzerland 2014

Need to Combine Numerical Characteristics, and the Emergence of Polynomial Aggregation Formulas.

- In decision making, we need to combine utilities $u_1, \ldots, u_n$ of different participants. Nobelist Josh Nash showed that reasonable conditions lead to $u = u_1 \cdot \ldots \cdot u_n$; see, e.g., [1, 2].

- In space-time geometry, we need to combine coordinates x_i into a metric; reasonable conditions lead to polynomial metrics such as Minkowski metric in which

$$s^2 = c^2 \cdot (x_0 - x_0')^2 - (x_0 - x_0')^2 - (x_1 - x_1')^2 - (x_2 - x_2')^2 - (x_3 - x_3')^2$$

  and of a more general Riemann metric where $ds^2 = \sum_{i,j} g_{ij} \cdot dx^i \cdot dx^j$.

- In fuzzy logic and soft constraints, we must combine degrees of certainty d_i in A_i into a degree d for $A_1 \,\&\, A_2$; reasonable conditions lead to polynomial functions like $d = d_1 \cdot d_2$.

In Mathematical Terms, Polynomial Formulas Are Tensor-Related. In mathematical terms, a general polynomial dependence

$$f(x_1, \ldots, x_n) = f_0 + \sum_{i=1}^{n} f_i \cdot x_i + \sum_{i=1}^{n} \sum_{j=1}^{n} f_{ij} \cdot x_i \cdot x_j + \sum_{i=1}^{n} \sum_{j=1}^{n} \sum_{k=1}^{n} f_{ijk} \cdot x_i \cdot x_j \cdot x_k + \ldots$$

means that to describe this dependence, we need a finite collection of tensors f_0, f_i, f_{ij}, f_{ijk}, $\ldots$, of different arity.

Towards a General Justification of Polynomial (Tensor) Formulas. The fact that similar polynomials appear in different application areas indicates that there is a common reason behind them. In this paper, we provide such a general justification.

We want to find a finite-parametric class F of analytical functions $f(x_1, \ldots, x_n)$ approximating the actual complex aggregation. It is reasonable to require that this class F be invariant with respect to addition and multiplication by a constant, i.e., that it is a (finite-dimensional) linear space of functions.

The invariance with respect to multiplication by a constant corresponds to the fact that the aggregated quantity is usually defined only modulo the choice of a measuring unit. If we replace the original measuring unit by a one which is λ times smaller, then all the numerical values get multiplied by this factor λ: $f(x_1, \ldots, x_n)$ is replaced with $\lambda \cdot f(x_1, \ldots, x_n)$.

Similarly, in all three areas, the numerical values x_i are defined modulo the choice of a measuring unit. If we replace the original measuring unit by a one which is λ times smaller, then all the numerical values get multiplied by this factor λ: x_i is replaced with $\lambda \cdot x_i$. It is therefore reasonable to also require that the finite-dimensional linear space F be invariant with respect to such re-scalings, i.e., if $f(x_1, \ldots, x_n) \in F$, then for every $\lambda > 0$, the function

$$f_\lambda(x_1, \ldots, x_n) \stackrel{\text{def}}{=} f(\lambda \cdot x_1, \ldots, \lambda \cdot x_n)$$

also belongs to the family F.

Under this requirement, we prove that all elements of F are polynomials.

Definition 1. *Let n be an arbitrary integer. We say that a finite-dimensional linear space F of analytical functions of n variables is* scale-invariant *if for every $f \in F$ and for every $\lambda > 0$, the function*

$$f_\lambda(x_1, \ldots, x_n) \overset{\text{def}}{=} f(\lambda \cdot x_1, \ldots, \lambda \cdot x_n)$$

also belongs to the family F.

Main Result. *For every scale-invariant finite-dimensional linear space F of analytical functions, every element $f \in F$ is a polynomial.*

Proof. Let F be a scale-invariant finite-dimensional linear space F of analytical functions, and let $f(x_1, \ldots, x_n)$ be a function from this family F.

By definition, an analytical function $f(x_1, \ldots, x_n)$ is an infinite series consisting of monomials $m(x_1, \ldots, x_n)$ of the type

$$a_{i_1 \ldots i_n} \cdot x_1^{i_1} \cdot \ldots \cdot x_n^{i_n}.$$

For each such term, by its *total order*, we will understand the sum $i_1 + \ldots + i_n$. The meaning of this total order is simple: if we multiply each input of this monomial by λ, then the value of the monomial is multiplied by λ^k:

$$m(\lambda \cdot x_1, \ldots \lambda \cdot x_n) = a_{i_1 \ldots i_n} \cdot (\lambda \cdot x_1)^{i_1} \cdot \ldots \cdot (\lambda \cdot x_n)^{i_n} =$$

$$\lambda^{i_1 + \ldots + i_n} \cdot a_{i_1 \ldots i_n} \cdot x_1^{i_1} \cdot \ldots \cdot x_n^{i_n} = \lambda^k \cdot m(x_1, \ldots, x_n).$$

For each order k, there are finitely many possible combinations of integers $i_1, \ldots, i_n$ for which $i_1 + \ldots + i_n = k$, so there are finitely many possible monomials of this order. Let $P_k(x_1, \ldots, x_n)$ denote the sum of all the monomials of order k from the series describing the function $f(x_1, \ldots, x_n)$. Then, we have

$$f(x_1, \ldots, x_n) = P_0 + P_1(x_1, \ldots, x_n) + P_2(x_1, x_2, \ldots, x_n) + \ldots$$

Some of these terms may be zeros – if the original expansion has no monomials of the corresponding order. Let k_0 be the first index for which the term $P_{k_0}(x_1, \ldots, x_n)$ is not identically 0. Then,

$$f(x_1, \ldots, x_n) = P_{k_0}(x_1, \ldots, x_n) + P_{k_0+1}(x_1, x_2, \ldots, x_n) + \ldots$$

Since the family F is scale-invariant, it also contains the function

$$f_\lambda(x_1, \ldots, x_n) = f(\lambda \cdot x_1, \ldots, \lambda \cdot x_n).$$

At this re-scaling, each term P_k is multiplied by λ^k; thus, we get

$$f_\lambda(x_1, \ldots, x_n) = \lambda^{k_0} \cdot P_{k_0}(x_1, \ldots, x_n) + \lambda^{k_0+1} \cdot P_{k_0+1}(x_1, x_2, \ldots, x_n) + \ldots$$

Since F is a linear space, it also contains a function

$$\lambda^{-k_0} \cdot f_\lambda(x_1, \ldots, x_n) = P_{k_0}(x_1, \ldots, x_n) + \lambda \cdot P_{k_0+1}(x_1, x_2, \ldots, x_n) + \ldots$$

Since F is finite-dimensional, it is closed under turning to a limit. In the limit $\lambda \to 0$, we conclude that the term $P_{k_0}(x_1, \ldots, x_n)$ also belongs to the family F.

Since F is a linear space, this means that the difference

$$f(x_1, \ldots, x_n) - P_{k_0}(x_1, \ldots, x_n) =$$

$$P_{k_0+1}(x_1, x_2, \ldots, x_n) + P_{k_0+2}(x_1, x_2, \ldots, x_n) + \ldots$$

also belongs to F. If we denote, by k_1, the first index $k_1 > k_0$ for which the term $P_{k_1}(x_1, \ldots, x_n)$ is not identically 0, then we can similarly conclude that this term $P_{k_1}(x_1, \ldots, x_n)$ also belongs to the family F, etc.

We can therefore conclude that for every index k for which term $P_k(x_1, \ldots, x_n)$ is not identically 0, this term $P_k(x_1, \ldots, x_n)$ also belongs to the family F.

Monomials of different total order are linearly independent. Thus, if there were infinitely many non-zero terms P_k in the expansion of the function $f(x_1, \ldots, x_n)$, we would have infinitely many linearly independent function in the family F – which contradicts to our assumption that the family F is a finite-dimensional linear space.

So, in the expansion of the function $f(x_1, \ldots, x_n)$, there are only finitely many non-zero terms. Hence, the function $f(x_1, \ldots, x_n)$ is a sum of finitely many monomials – i.e., a polynomial.

The statement is proven.

Acknowledgments. This work was supported in part by the National Science Foundation grants HRD-0734825 and DUE-0926721, by Grant 1 T36 GM078000-01 from the National Institutes of Health, by Grant MSM 6198898701 from MŠMT of Czech Republic, and by Grant 5015 "Application of fuzzy logic with operators in the knowledge based systems" from the Science and Technology Centre in Ukraine (STCU), funded by European Union.

References

1. Luce, R.D., Raiffa, R.: Games and decisions: introduction and critical survey. Dover, New York (1989)
2. Nguyen, H.T., Kosheleva, O., Kreinovich, V.: Decision Making Beyond Arrow's Impossibility Theorem. International Journal of Intelligent Systems 24(1), 27–47 (2009)
3. Nguyen, H.T., Walker, E.A.: A First Course in Fuzzy Logic. Chapman & Hall/CRC Press, Boca Raton (2006)

Adding Constraints –
A (Seemingly Counterintuitive but) Useful Heuristic in Solving Difficult Problems

Olga Kosheleva, Martine Ceberio, and Vladik Kreinovich

University of Texas at El Paso,
El Paso, TX 79968, USA
{olgak,mceberio,vladik}@utep.edu

Abstract. Intuitively, the more constraints we impose on a problem, the more difficult it is to solve it. However, in practice, difficult-to-solve problems sometimes get solved when we impose additional constraints and thus, make the problems seemingly more complex. In this methodological paper, we explain this seemingly counter-intuitive phenomenon, and we show that, dues to this explanation, additional constraints can serve as a useful heuristic in solving difficult problems.

Keywords: constraints, algorithmic problems, heuristics.

Commonsense Intuition: The More Constraints, the More Difficult the Problem. Intuitively, the more constraints we impose on a problem, the more difficult it is to solve it.

For example, if a university has a vacant position of a lecturer in Computer Science Department, and we want to hire a person with a PhD in Computer Science to teach the corresponding classes, then this hiring is a reasonably easy task. However, once we impose constraints: that the person has several years of teaching experience at similar schools and has good evaluations to show for this experience, that this person's research is in the area close to the classes that he or she needs to teach, etc., then hiring becomes a more and more complicated task.

If a person coming to a conference is looking for a hotel to stay, this is usually an easy problem to solve. But once you start adding constraints on how far this hotel is from the conference site, how expensive it is, how noisy it is, etc., the problems becomes difficult to solve.

Similarly, in numerical computations, unconstrained optimization problems are usually reasonably straightforward to solve, but once we add constraints, the problems often become much more difficult.

Sometimes Constraints Help: A Seemingly Counterintuitive Phenomenon. In practice, difficult-to-solve problems sometimes get solved when we impose additional constraints and thus, make the problems seemingly more complex.

Sometimes this easiness to solve is easy to explain. For example, when a traveler prefers a certain hotel chain, and make this chain's brand name a constraint,

M. Ceberio and V. Kreinovich (eds.), *Constraint Programming and Decision Making,*   79
Studies in Computational Intelligence 539,
DOI: 10.1007/978-3-319-04280-0_10, © Springer International Publishing Switzerland 2014

then making reservations in a small town is usually not a difficult problem to solve, because in this town, there is usually only one hotel from this chain.

However, in other cases, the resulting easiness-to-solve is not so easy to explain.

Many such examples come from mathematicians solving practical problems. For example, in application problems, mathematicians often aim for an optimal control or an optimal design. To a practitioner, this desire for the exact optimum may seem like a waste of time. Yes, it is desirable to find an engineering design with the smallest cost under the given constraints – or, vice versa, with the best performance under the given cost constraints – but since we can predict the actual consequences of each design only approximately, wasting time to exactly optimize the approximately optimize the approximately known function does not seem to make sense. If we only know the objective function $f(x)$ with accuracy $\varepsilon > 0$ (e.g., 0.1), then once we are within ε of the maximum, we can as well stop.

In some cases, it is sufficient to simply satisfy some constraint $f(x) \geq f_0$ for some value f_0. However, from the algorithmic viewpoint, often, the best way to solve this problem is to find the maximum of the function $f(x)$ on a given domain – by equating partial derivatives of $f(x)$ to 0. If there is a value x for which $f(x) \geq f_0$, then definitely $\max_y f(y) \geq f_0$, so the place x where the function $f(y)$ attains its maximum satisfies the desired constraint. In other words, by imposing an additional constraint – that not only $f(x) \geq f_0$, but also that $f(x) = \max_y f(y)$ – we make the problem easier to solve.

In theoretical mathematics, a challenging hypothesis often becomes proven when instead of simply looking for its proof, we look for proofs that can be applied to other cases as well – in other words, when we apply an additional constraint of generalizability; see, e.g., [16] and references therein.

Similarly, interesting results about a physical system become proven in the realm of rigorous mathematics, while, due to the approximate character of the model, arguments on the physical level of rigor would be (and often are) sufficient.

In engineering and science, often, problems get solved when someone starts looking not just for a solution but for a solution that satisfies additional constraints of symmetry, beauty, etc. – or when a physicist looks for a physical theory that fits his philosophical view of the world; a large number of examples how the search for a beautiful solution helped many famous mathematicians and physicists – including Bolzmann and Einstein – are described in [8].

In software design, at first glance, additional constraints imposed by software engineering – like the need to have comments, the need to have simple modules, etc. – seem to make a problem more complicated, but in reality, complex designs often become possible only after all these constraints are imposed.

This phenomenon extends to informal problems as well. For example, in art, many great objects have been designed within strict requirements on shape, form, etc. – under the constraints of a specific reasonable regulated style of music, ballet, poetry, painting, while free-form art while seemingly simpler and less restrictive, does not always lead to more impressive art objects. Some people find personal happiness when accepting well-regulated life rules – e.g., within a traditional religious community – while they could not find personal happiness in their earlier freer life.

How can we explain this seemingly counter-intuitive phenomenon?

Analysis of the Problem. By definition, when we impose an additional constraint, this means that some alternatives which were originally solutions to the problem, stop being such solutions – since we impose extra constraints, constraints that are not always satisfied by all original solutions.

Thus, the effect of adding a constraint is that the number of solution decreases. At the extreme, when we have added the largest possible number of constraints, we get a unique solution.

It turns out that this indeed explains why adding constraints can make the problems easier.

Related Known Results: The Fewer Solutions, the Easier to Solve the Problem. Many numerical problems are, in general, algorithmically undecidable: for example, no algorithm can always find a solution to an algorithmically defined system of equation or find a location of the maximum of an algorithmically defined function; see, e.g., [1, 2, 4–6, 17, 18, 22].

The proofs of most algorithmic non-computability results essentially use functions which have several maxima and/or equations which have several solutions. It turned out that this is not an accident: uniqueness actually implies algorithmic computability. Such a result was first proven in [19], where an algorithm was designed that inputs a constructive function of one or several real variables on a bounded set that attains its maximum on this set at exactly one point – and computes this global maximum point. In [20], this result was to constructive functions on general constructive compact spaces.

In [12, 14], this result was applied to design many algorithms: from optimal approximation of functions to designing a convex body from its metric to constructive a shortest path in a curved space to designing a Riemannian space most tightly enclosing unit spheres in a given Finsler space [7]. Several efficient algorithms based on uniqueness have been described in [9–11].

On the other hand, it was proven that a general algorithm is not possible for functions that have exactly two global maxima or systems that have exactly two solutions; see, e.g., [12–15, 17].

Moreover, there are results showing that for every m, problems with exactly m solutions are, in general, more computationally difficult than problems with $m - 1$ solutions; see, e.g., [21].

Resulting Recommendation. The above discussion leads to the following seemingly counter-intuitive recommendation: If a problem turns out to be too complex to solve, maybe a good heuristic is to add constraints and make it more complex.

For example, if the problem that we have difficulty solving is an applied mathematical problem, based on an approximate description of reality, maybe a good idea is not to simplify this problem but rather to make it more realistic. This recommendation may sound counter-intuitive, but applied mathematicians know that often, learning more about the physical or engineering problem helps to solve it.

This can also be applied to education. If students have a hard time solving a class of problems, maybe a good idea is not to make these problems easier, but to make them more complex. Again, at first glance, this recommendation may sound counter-intuitive, but in pedagogy, it is a known fact: if a school is failing, the solution is usually not to make classes easier – this will lead to a further decline in knowledge. Anecdotal evidence shows that a turnaround happens when a new teacher starts giving students more complex more challenging problems – and this boosts their knowledge.

This recommendation is in line with a general American idea – that to be satisfying, the job, among other things, must be a challenge.

Caution. Of course, it is important not to introduce so many constraints that the problem simply stops having solutions at all. Since it is difficult to guess which level of constraints will lead to inconsistency, it may be a good idea to simultaneously several different versions of the original problem, with different number of constraints added – this way, we will hopefully be able to successfully solve one of them.

Acknowledgments. This work was supported in part by the National Science Foundation grants HRD-0734825 and DUE-0926721 and by Grant 1 T36 GM078000-01 from the National Institutes of Health.

References

1. Aberth, O.: Precise Numerical Analysis Using C++. Academic Press, New York (1998)
2. Beeson, M.J.: Foundations of Constructive Mathematics. Springer, New York (1985)
3. Bishop, E.: Foundations of Constructive Analysis. McGraw-Hill, New York (1967)
4. Bishop, E., Bridges, D.S.: Constructive Analysis. Springer, New York (1985)
5. Bridges, D.S.: Constructive Functional Analysis. Pitman, London (1979)
6. Bridges, D.S., Via, S.L.: Techniques of Constructive Analysis. Springer, New York (2006)
7. Busemann, H.: The Geometry of Geodesics. Dover Publ., New York (2005)
8. Chandrasekhar, S.: Beauty and the quest for beauty in science. Physics Today 32(7), 25–30 (1979); reprinted in 62(12), 57–62 (2010)
9. Kohlenbach, U.: Theorie der majorisierbaren und stetigen Funktionale und ihre Anwendung bei der Extraktion von Schranken aus inkonstruktiven Beweisen: Effektive Eindeutigkeitsmodule bei besten Approximationen aus ineffektiven Eindeutigkeitsbeweisen. Ph.D. Dissertation, Frankfurt am Main (1990) (in German)
10. Kohlenbach, U.: Effective moduli from ineffective uniqueness proofs. An unwinding of de La Vallée Poussin's proof for Chebycheff approximation. Annals for Pure and Applied Logic 64(1), 27–94 (1993)

11. Kohlenbach, U.: Applied Proof Theory: Proof Interpretations and their Use in Mathematics. Springer, Heidelberg (2008)
12. Kreinovich, V.: Uniqueness implies algorithmic computability. In: Proceedings of the 4th Student Mathematical Conference, pp. 19–21. Leningrad University, Leningrad (1975) (in Russian)
13. Kreinovich, V.: Reviewer's remarks in a review of Bridges, D.S.: Constrictive functional analysis. Pitman, London (1979); Zentralblatt für Mathematik 401, 22–24 (1979)
14. Kreinovich, V.: Categories of space-time models. Ph.D. dissertation, Novosibirsk, Soviet Academy of Sciences, Siberian Branch, Institute of Mathematics (1979) (in Russian)
15. Kreinovich, V.: Physics-motivated ideas for extracting efficient bounds (and algorithms) from classical proofs: beyond local compactness, beyond uniqueness. In: Abstracts of the Conference on the Methods of Proof Theory in Mathematics, June 3-10, p. 8. Max-Planck Institut für Mathematik, Bonn (2007)
16. Kreinovich, V.: Any (true) statement can be generalized so that it becomes trivial: a simple formalization of D. K. Faddeev's belief. Applied Mathematical Sciences 47, 2343–2347 (2009)
17. Kreinovich, V., Lakeyev, A., Rohn, J., Kahl, P.: Computational complexity and feasibility of data processing and interval computations. Kluwer, Dordrecht (1998)
18. Kushner, B.A.: Lectures on Constructive Mathematical Analysis. Amer. Math. Soc. Providence, Rhode Island (1984)
19. Lacombe, D.: Les ensembles récursivement ouvert ou fermés, et leurs applications à l'analyse récurslve. Compt. Rend. 245(13), 1040–1043 (1957)
20. Lifschitz, V.A.: Investigation of constructive functions by the method of fillings. J. Soviet Math. 1, 41–47 (1973)
21. Longpré, L., Kreinovich, V., Gasarch, W., Walster, G.W.: m Solutions Good, $m-1$ Solutions Better. Applied Math. Sciences 2(5), 223–239 (2008)
22. Pour-El, M., Richards, J.: Computability in Analysis and Physics. Springer, New York (1989)

Under Physics-Motivated Constraints, Generally-Non-Algorithmic Computational Problems become Algorithmically Solvable

Vladik Kreinovich

Department of Computer Science,
University of Texas at El Paso,
El Paso, TX 79968, USA
vladik@utep.edu
http://www.cs.utep.edu/vladik

Abstract. It is well known that many computational problems are, in general, not algorithmically solvable: e.g., it is not possible to algorithmically decide whether two computable real numbers are equal, and it is not possible to compute the roots of a computable function. We propose to constraint such operations to certain "sets of typical elements" or "sets of random elements".

In our previous papers, we proposed (and analyzed) physics-motivated definitions for these notions. In short, a set T is a *set of typical elements* if for every definable sequences of sets A_n with $A_n \supseteq A_{n+1}$ and $\bigcap_n A_n = \emptyset$, there exists an N for which $A_N \cap T = \emptyset$; the definition of a *set of random elements* with respect to a probability measure P is similar, with the condition $\bigcap_n A_n = \emptyset$ replaced by a more general condition $\lim_n P(A_n) = 0$.

In this paper, we show that if we restrict computations to such typical or random elements, then problems which are non-computable in the general case – like comparing real numbers or finding the roots of a computable function – become computable.

Keywords: constraints, computable problems, random elements, typical elements.

Physically Meaningful Computations with Real Numbers: A Brief Reminder. In practice, many quantities such as weight, speed, etc., are characterized by real numbers. To get information about the corresponding value x, we perform measurements. Measurements are never absolute accurate. As a result of each measurement, we get a measurement result $\tilde{x}$; for each measurement, we usually also know the upper bound Δ on the (absolute value of) the measurement error

$$\Delta x \overset{\text{def}}{=} \tilde{x} - x: \; |x - \tilde{x}| \leq \Delta.$$

To fully characterize a value x, we must measure it with a higher and higher accuracy. As a result, when we perform measurements with accuracy 2^{-n} with $n = 0, 1, \ldots$, we get a sequence of rational numbers r_n for which $|x - r_n| \leq 2^{-n}$.

M. Ceberio and V. Kreinovich (eds.), *Constraint Programming and Decision Making*,
Studies in Computational Intelligence 539,
DOI: 10.1007/978-3-319-04280-0_11, © Springer International Publishing Switzerland 2014

From the algorithmic viewpoint, we can view this sequence as an oracle that, given an integer n, returns a rational number r_n. Such sequences represent real numbers in computable analysis; see, e.g., [9, 10].

First Negative Result. In computable analysis, several negative results are known. For example, it is known that no algorithm is possible that, given two numbers x and y, would check whether these numbers are equal or not.

Computable Functions and Relative Negative Results. Similarly, we can define a function $f(x)$ from real numbers to real numbers as a mapping that, given an integer n, a rational number x_m and its accuracy m, produces either a message that this information is insufficient, or a rational number y_n which is 2^{-n}-close to all the values $f(x)$ for $d(x, x_m) \leq 2^{-m}$ – and for which, for every x and for each desired accuracy n, there is an m for which a rational number y_n is produced. We can also define a computable function $f(x_1, \ldots, x_k)$ of several real variables (and, even more generally, a function on a computable compact).

Several negative results are known about computable functions as well. For example,

- while there is an algorithm that, given a function $f(x)$ on a computable compact set K (e.g., on a box $[\underline{x}_1, \overline{x}_1] \times \ldots \times [\underline{x}_k, \overline{x}_k]$ in k-dimensional space), produces the values $\max\{f(x) : x \in K\}$,
- no algorithm is possible that would always return a point x at which this maximum is attained (and similarly, with minimum).

From the Physicists' Viewpoint, These Negative Results Seem Rather Theoretical. From the purely mathematical viewpoint, if two quantities coincide up to 13 digits, they may still turn to be different: for example, they may be 1 and $1 + 10^{-100}$.

However, in the physics practice, if two quantities coincide up to a very high accuracy, it is a good indication that they are actually equal. This is how physical theories are confirmed: if an experimentally observed value of a quantity turned out to be very close to the value predicted based on a theory, this means that this theory is (triumphantly) true. This is, for example, how General Relativity has been confirmed.

This is how discoveries are often made: for example, when it turned out the speed of the waves described by Maxwell equations of electrodynamics is very close to the observed speed of light c, this led physicists to realize that light is formed of electromagnetic waves.

How Physicists Argue. A typical physicist argument is that while numbers like $1 + 10^{-100}$ (or $c \cdot (1 + 10^{-100})$) are, in principle, possible, they are abnormal (not typical).

When a physicist argues that second order terms like $a \cdot \Delta x^2$ of the Taylor expansion can be ignored in some approximate computations because Δx is small, the argument is that

- while abnormally high values of a (e.g., $a = 10^{40}$) are mathematically possible,
- typical (= not abnormal) values appearing in physical equations are usually of reasonable size.

How to Formalize the Physicist's Intuition of Typical (Not Abnormal). A formalization of this intuition was proposed and analyzed in [1–7]. Its main idea is as follows. To some physicist, all the values of a coefficient a above 10 are abnormal. To another one, who is more cautious, all the values above 10 000 are abnormal. Yet another physicist may have another threshold above which everything is abnormal. However, for every physicist, there is a value n such that all value above n are abnormal.

This argument can be generalized as a following property of the set T of all typical elements. Suppose that we have a monotonically decreasing sequence of sets $A_1 \supseteq A_2 \supseteq \ldots$ for which $\bigcap_n A_n = \emptyset$ (in the above example, A_n is the set of all numbers $\geq n$). Then, there exists an integer N for which $T \cap A_N = \emptyset$.

We thus say that T is a *set of typical elements* if for every definable decreasing sequence $\{A_n\}$ for which $\bigcap_n A_n = \emptyset$, there exists an N for which $T \cap A_N = \emptyset$.

Comment. Of course, to make this definition precise, we must restrict definability to a *subset* of properties, so that the resulting notion of definability will be defined in ZFC itself (or in whatever language we use); for details, see, e.g., [3].

Relation to Randomness. The above notion of typicality is related to the randomness. Indeed, a usual definition of a random sequence (see, e.g., [8]) is based on the idea that a sequence is random if it satisfies all the probability laws – like the law of large numbers, the central limit theorem, etc. A probability law is then described as a definable property that is satisfied with probability 1, i.e., as a complement to a definable set S of probability measure 0 ($P(S) = 0$). Thus, we can say that a sequence is random if it does not belong to any definable set of measure 0. (If we use different languages to formalize the notion "definable", we get different versions of Kolmogorov-Martin-Löf randomness.)

Informally, this definition means that (definable) events with probability 0 cannot happen. In practice, physicists also assume that events with a *very small* probability cannot happen. It is not possible to formalize this idea by simply setting a threshold $p_0 > 0$ below which events are not possible – since then, for N for which $2^{-N} < p_0$, no sequence of N heads or tails would be possible at all. However, we know that for each monotonic sequence of properties A_n with $\lim p(A_n) = 0$ (e.g., $A_n =$ "we can get first n heads"), there exists an N above which a truly random sequence cannot belong to A_N. In [1–7], we thus propose to describe a set R as a set of random elements if it satisfies the following property: for every definable decreasing sequence $\{A_n\}$ for which $\lim P(A_n) = 0$, there exists an N for which $R \cap A_N = \emptyset$.

It turns out that properties of T and R are related:

- every set of random elements is also a set of typical elements, and
- for every set of typical elements T, the difference $T - R_K$, where R_K is the set of the elements random in the usual Kolmogorov-Martin-Löf sense, is a set of random elements [2].

Physically Interesting Consequences of These Definitions. These definitions have useful consequences [1–7].

For example, when the universal set X is a metric space, both sets T and R are *pre-compact* – with the consequence that all inverse problems become well-defined: for any 1-1 continuous function $f : X \to X$, the restriction of the inverse function to T is also continuous. This means that, in contrast to ill-defined problem, if we perform measurements accurately enough, we can reconstruct the state of the system with any desired accuracy.

Another example is a justification of physical induction: crudely speaking, there exists an N such that if for a typical sequence, a property is satisfied in the first N experiments, then it is satisfied always.

New Results: When We Restrict Ourselves to Typical Elements, Algorithms become Possible. In this paper, we analyze the computability consequences of the above definitions. Specifically, we show that most negative results of computability analysis disappear if we restrict ourselves to typical elements.

For example, for every set of typical pairs of real numbers $T \subseteq \mathbb{R}^2$, there exists an algorithm, that, given real numbers $(x, y) \in T$, decides whether $x = y$ or not. To prove it, consider a decreasing sequence of definable sets

$$A_n = \{(x, y) : 0 < d(x, y) < 2^{-n}\}.$$

By definition of T, there exists an N such that $A_N \cap T = \emptyset$. Thus, if we compute $d(x, y)$ with accuracy $2^{-(N+1)}$ and get a value $< 2^{-N}$, this means that $x = y$ – otherwise $x \neq y$.

Similar (but somewhat more complex) arguments lead to

- an algorithm that, given a typical function $f(x)$ on a computable compact K, computes a value x at which $f(x)$ attains its maximum,
- an algorithm that, given a typical function $f(x)$ on a computable compact K that attains a 0 value somewhere on K, computes a value x at which $f(x) = 0$,
- etc.

Acknowledgments. This work was supported in part by the National Science Foundation grants HRD-0734825 and DUE-0926721, by Grant 1 T36 GM078000-01 from the National Institutes of Health, and by Grant MSM 6198898701 from MŠMT of Czech Republic.

References

1. Finkelstein, A.M., Kreinovich, V.: Impossibility of hardly possible events: physical consequences. In: Abstracts of the 8th International Congress on Logic, Methodology, and Philosophy of Science, 1987, Moscow, vol. 5(2), pp. 23–25 (1987)
2. Kreinovich, V.: Toward formalizing non-monotonic reasoning in physics: the use of Kolmogorov complexity. Revista Iberoamericana de Inteligencia Artificial 41, 4–20 (2009)
3. Kreinovich, V., Finkelstein, A.M.: Towards applying computational complexity to foundations of physics. Notes of Mathematical Seminars of St. Petersburg Department of Steklov Institute of Mathematics 316, 63–110 (2004); reprinted in Journal of Mathematical Sciences 134(5), 2358–2382 (2006)
4. Kreinovich, V., Kunin, I.A.: Kolmogorov complexity and chaotic phenomena. International Journal of Engineering Science 41(3), 483–493 (2003)
5. Kreinovich, V., Kunin, I.A.: Kolmogorov complexity: how a paradigm motivated by foundations of physics can be applied in robust control. In: Fradkov, A.L., Churilov, A.N. (eds.) Proceedings of the International Conference "Physics and Control" PhysCon 2003, Saint-Petersburg, Russia, August 20-22, pp. 88–93 (2003)
6. Kreinovich, V., Kunin, I.A.: Application of Kolmogorov complexity to advanced problems in mechanics. In: Proceedings of the Advanced Problems in Mechanics Conference APM 2004, St. Petersburg, Russia, June 24-July 1, pp. 241–245 (2004)
7. Kreinovich, V., Longpré, L., Koshelev, M.: Kolmogorov complexity, statistical regularization of inverse problems, and Birkhoff's formalization of beauty. In: Mohamad-Djafari, A. (ed.) Bayesian Inference for Inverse Problems, Proceedings of the SPIE/International Society for Optical Engineering, San Diego, California, vol. 3459, pp. 159–170 (1998)
8. Li, M., Vitanyi, P.: An Introduction to Kolmogorov Complexity and Its Applications. Springer (2008)
9. Pour-El, M.B., Richards, J.I.: Computability in Analysis and Physics. Springer, Berlin (1989)
10. Weihrauch, K.: Computable Analysis. Springer, Berlin (2000)

Constraint-Related Reinterpretation of Fundamental Physical Equations Can Serve as a Built-In Regularization

Vladik Kreinovich[1], Juan Ferret[2], and Martine Ceberio[1]

[1] Department of Computer Science
[2] Department of Philosophy,
University of Texas at El Paso, El Paso, TX 79968, USA
{vladik,jferret,mceberio}@utep.edu

Abstract. Many traditional physical problems are known to be *ill-defined*: a tiny change in the initial condition can lead to drastic changes in the resulting solutions. To solve this problem, practitioners *regularize* these problem, i.e., impose explicit constraints on possible solutions (e.g., constraints on the squares of gradients). Applying the Lagrange multiplier techniques to the corresponding constrained optimization problems is equivalent to adding terms proportional to squares of gradients to the corresponding optimized functionals. It turns out that many optimized functionals of fundamental physics already have such squares-of-gradients terms. We therefore propose to re-interpret these equations – by claiming that they come not, as it is usually assumed, from unconstrained optimization, but rather from a constrained optimization, with squares-of-gradients constrains. With this re-interpretation, the physical equations remain the same – but now we have a built-in regularization; we do not need to worry about ill-defined solutions anymore.

Keywords: constraints, fundamental physics, regularization, ill-defined problems.

1 Formulation of the Problem

Optimization Reformulation of Physical Equations. Traditionally, laws of physics have been described in terms of differential equations. However, in the 19th century, it turned out that these equations can be reformulated as optimization problems: the actual field is the one that minimizes the corresponding functional (called *action S*). This optimization approach is very useful in many applications (see, e.g., [1]) since there are many efficient algorithms for solving optimization problems.

Decision Making and Control: Ideal Situation. In decision making and control applications, in principle, we can similarly predict the result of different decisions, different control strategies. Thus, we can select the decision (or the control strategy) that leads to the most favorable result.

M. Ceberio and V. Kreinovich (eds.), *Constraint Programming and Decision Making,*   91
Studies in Computational Intelligence 539,
DOI: 10.1007/978-3-319-04280-0_12, © Springer International Publishing Switzerland 2014

Real-Life Prediction: Limitations. In practice, however, the situation is not so simple. The main problem is that all measurements are only approximate. Even for the most accurate measurements, the measured values of the initial conditions are slightly different from the actual values.

Most prediction problems are *ill-defined* in the sense that small deviations in the initial conditions can cause arbitrary large deviations in the predicted values.

Limitations: Example. One of the main reasons why the prediction problem is ill-defined is that no matter how small a sensor is, it always has a finite size. As a result, the sensor does not produce the value $f(x)$ of the measured field f exactly at a given spatial location x; the sensor always captures the "average" value of a signal over a certain neighborhood of the point x – the neighborhood that is occupied by this sensor. Hence, field components with high spatial frequency $f(x) = f_0 \cdot \sin(\omega \cdot x)$ (with large ω) are averaged out and thus, not affected by the measurement result. Therefore, in addition to the measured field $f(x)$, the same measurement result could be produced by a different field $f(x) + f_0 \cdot \sin(\omega \cdot x)$. For many differential equations, future predictions based on this new field can be drastically different from the predictions corresponding to the original field $f(x)$.

How This Problem Is Solved Now. To solve the problem, practitioners use *regularization*, i.e., in effect, restrict themselves to the class of solutions that satisfies a certain constraint; see, e.g., [5]. For example, for fields $f(x)$, typical constraints include bounds on the values $\int f^2 \, dx$ and $\int f_{,i} \cdot f^{,i} \, dx$, where $\int F \, dx$ means integration over space-time (or, for static problems, over space), $f_{,i} \overset{\text{def}}{=} \dfrac{\partial f}{\partial x_i}$, and an expression $f_{,i} \cdot f^{,i}$ means summation over all coordinates i.

By imposing bounds on the derivatives, we thus restrict the possibility of high-frequency components of the type $f_0 \cdot \sin(\omega \cdot x)$ and thus, make the problem well-defined.

Limitations. The main limitation of different regularization techniques is that the bounds on the derivatives are introduced *ad hoc*, they do not follow from the physics, and different bounds lead to different solutions.

There is a whole art of selecting an appropriate regularization techniques, and, once a technique is selected, of selecting an appropriate parameter. It is desirable to come up with a more algorithmic way to making the equations well-defined.

2   Main Idea

A Mathematical Reminder: How to Optimize Functionals (see, e.g., [2]) As we have mentioned, fundamental physical equations are described in terms of minimizing a functional called *action*. This functional usually has an integral form $S = \int L(f, f_{,i}) \, dx$; the corresponding function L is called a *Lagrangian*.

The main idea behind minimizing such functional is similar to the idea of minimizing functions. For functions $f(x_1, \ldots, x_n)$, optima occur when all the partial derivatives are 0s. Similarly, for a functional, an optimum occurs if the *functional derivative* is 0:

$$\frac{\delta L}{\delta f} \stackrel{\text{def}}{=} \frac{\partial L}{\partial f} \cdot \Delta f - \frac{\partial}{\partial x_i} \left(\frac{\partial L}{\partial f_i} \right) = 0.$$

This is how usual differential equations are derived from the optimization reformulation of the corresponding physical theories.

A Mathematical Reminder: How Constraints Are Currently Taken into Account? When we optimize a functional, e.g., $\int f^2 \, dx$, under a constraint such as

$$\int f_{,i} \cdot f^{,i} \, dx \leq \Delta,$$

then, from the mathematical viewpoint, there are two options:

- It is possible that the optimum of the functional is attained strictly inside the area defined by the constraints. In the above example, it means that the optimum is attained when $\int f_{,i} \cdot f^{,i} \, dx < \Delta$. In this case, all the (functional) derivatives of the original functional are equal to 0. So, in effect, in this case, we have regular physical equations – unaffected by constraints. We have already mentioned that in this case, we often get ill-defined solutions.
- The case when the constrains do affect the solutions is when that the optimum of the functional is attained on the border of the area defined by the constraints. In the above example, it means that the optimum is attained when $\int f_{,i} \cdot f^{,i} \, dx = \Delta$.

Therefore, in cases when constrains are important to impose (and do not just come satisfied "for free" already for the usual solution), the inequality-type constraints are equivalent to equality-type ones.

Optimization under such equality constraints is done by using the usual Lagrange multiplier approach: optimizing a functional F under a constraint $G = g_0$ (i.e., equivalently, $G - g_0 = 0$) is equivalent, for an appropriate real number λ, to an unconstraint optimization of an auxiliary functional $F + \lambda \cdot (G - g_0)$. The value λ must then be found from the constraint $G = g_0$.

In the above example, optimizing a functional $\int f^2 \, dx$ under a constraint $\int f_{,i} \cdot f^{,i} \, dx = \Delta$ is equivalent to an unconstrained optimization of the auxiliary functional

$$\int (f^2 + \lambda \cdot f_{,i} \cdot f^{,i}) \, dx.$$

Observation. The action functionals corresponding to fundamental physics theories already have a term proportional to $f_{,i} \cdot f^{,i}$ for a scalar field $f(x)$ or proportional to similar terms for more complex fields (vector, tensor, spinor, etc.)

Discussion. At present, this is what physicists are doing:

- They start with the (action) functionals $S = \int L\,dx$ corresponding to fundamental physical phenomena. These action functionals already have terms proportional to $f_{,i} \cdot f^{,i}$.
- Based on these action functionals, physicists derive the corresponding differential equations $\dfrac{\delta L}{\delta f} = 0$.
- A direct solution to the resulting differential equations is ill-defined (too much influenced by noise).
- Thus, instead of directly solving these equations, physicists *regularize* them, i.e., solve them under the constraints of the type $\int f_{,i} \cdot f^{,i}\,dx = \Delta$.

As we have mentioned, from the mathematical viewpoint, the regularization constrains are equivalent to adding terms of the type $f_{,i} \cdot f^{,i}$ to the corresponding Lagrangians. But these Lagrangians already have such terms! So, we arrive at a natural idea.

Idea. Traditionally, in fundamental physics, we assume that we have an *unconstrained* optimization $S = \int L\,dx \to$ min. A natural idea is to assume that in reality, the physical world corresponds to *constrained* optimization $F \to min$ under a constraint $G = g_0$ – and place terms like $f_{,i} \cdot f^{,i}$ into the constraint.

It Is Simply a Re-interpretation. At first glance, the above idea may sound like a sacrilege: a group of non-physicists challenge Einstein's equations? But we are *not* suggesting to change the equations, the differential equations – the only thing that we can check by observation – remain exactly the same. What we propose to change is the *interpretation* of these equations:

- Traditionally, these equations are interpreted via *unconstrained* optimization.
- We propose to interpret them via *constrained* optimization.

What Do We Gain? One might ask: if we are not proposing new equations, if we are not proposing any new physical theory, then what do we gain?

Our main gain is that we now have a built-in regularization. We do not need to worry about an additional outside regularization step anymore. We can not be sure that our problems are well-defined.

Possible Additional Gain. There may also be an additional gain, with respect to quantum versions of the fundamental physical theories. In contrast to the non-quantum field theory, in the quantum versions, if we impose the constraints, we do limit quantum solutions – because now, we are requiring the actual field to satisfy the additional constraint, while in the quantum case, all fields are possible (although with different probabilities). In quantum field theory, such absolute constraints are known as *super-selection* rules; see, e.g., [6]. It is known that such rules help to decrease divergence in quantum field theories (i.e., help them avoid these theories leading to meaningless infinite predictions); so maybe super-selection rules coming from our constrains will also be of similar help.

Possible Philosophical Meaning of Our Proposal. In addition to a pragmatic meaning (well-foundedness of the problem, possible decrease in divergence, etc.), our proposal may have a deeper philosophical meaning. To discuss such a meaning, let us consider the simplest possible case of a scalar field $f(x)$ corresponding to a particle of rest mass m. In the traditional field theory, its Lagrangian has the form $L = m^2 \cdot f^2 + f_{,i} \cdot f^{,i}$. For this theory, our proposal is, in effect, to make $\int f^2 \, dx$ an optimized function, and to introduce a constraint $\int f_{,i} \cdot f^{,i} \, dx = g_0$.

When we apply the Lagrange multiplier to this constrained optimization problem, we get the Lagrangian $L = f^2 + \lambda \cdot f_{,i} \cdot f^{,i}$ whose minimization is equivalent to minimizing $L' = \lambda^{-1} \cdot L = \lambda^{-1} \cdot f^2 + f_{,i} \cdot f^{,i}$. In other words, we recover the original Lagrangian, with $m^2 = \lambda^{-1}$. Now, in contrast to the traditional interpretation, the rest mass m is no longer the original fundamental parameter – it is a Lagrange multiplier that needs to be adjusted to fit the actual fundamental constant g_0 (which should be equal to $\int f_{,i} \cdot f^{,i} \, dx$).

Thus, the particle masses are no longer original fundamental constants – they depend on the fields in the rest of the world. This idea may sound somewhat heretic to a non-physicist, but it is very familiar to those who studied history of modern physics. This general philosophical idea – that all the properties like inertia, mass, etc. depend on the global configuration of the world – was promoted by a 19 century physicist Ernst Mach (see, e.g., [3]), and it was one of the main ideas that inspired Einstein to formulate his General Relativity theory [4], a theory in which what Einstein called *Mach's principle* is, to some extent, satisfied.

In other words, our idea may sound, at first glance, philosophically somewhat heretical, but it seems to be in line with Einstein's philosophical foundations for General Relativity.

Acknowledgments. This work was supported in part by the NSF grants HRD-0734825 and DUE-0926721, and by Grant 1 T36 GM078000-01 from NIH.

References

1. Feynman, R., Leighton, R., Sands, M.: The Feynman Lectures on Physics. Addison Wesley, Boston (2005)
2. Gelfand, I.M., Fomin, S.V.: Calculus of Variations. Dover Publ., New York (2000)
3. Mach, E.: The Science of Mechanics; a Critical and Historical Account of its Development. Open Court Pub. Co., LaSalle, Illinois (1960)
4. Misner, C.W., Thorne, K.S., Wheeler, J.A.: Gravitation. W.H. Freeman, New York (1973)
5. Tikhonov, A.N., Arsenin, V.Y.: Solutions of Ill-Posed Problems. W. H. Whinston & Sons, Washington, D.C. (1977)
6. Weinberg, S.: The Quantum Theory of Fields. Foundations, vol. 1. Cambridge University Press, Cambridge (1995)

Optimization of the Choquet Integral Using Genetic Algorithm

Tanja Magoč and François Modave*

Department of Computer Science,
University of Texas at El Paso,
500 W. University,
El Paso, Texas 79968, USA
{t.magoc,francois.modave}@gmail.com, francois.modave@ttuhsc.edu

Abstract. Decision making in an unavoidable part of out daily lives. Many decisions are straightforward, but others require careful consideration of each alternative and many attributes characterizing each alternative. If these attributes are mutually dependent, the Choquet integral is a technique often used for modeling the decision making problem. With a large number of attributes to consider, decision making becomes an optimization problem that requires huge computational resources in order to be solved exactly. Instead of using a large amount of these resources, heuristic techniques have been used to speed the computations and find a suboptimal decision. Yet, these heuristic methods could be improved to find better approximation with minimal increase in required computational resources. Genetic algorithm has been used in many situations as a heuristic optimization technique. In this paper, we present some modifications to the genetic algorithm that allow more precise optimization.

1 Introduction

We face situations in which we need to make decisions on daily basis. Some decisions are straightforward, while others are more complex and require more detailed analysis and usage of computational techniques. In these complex situations, the study of multi-criteria decision making is a helpful tool. While different techniques exist to solve a multi-criteria decision making problem, one of commonly used techniques when considering mutually dependent attributes of alternatives is the Choquet integral with respect to a 2-additive measure, which needs to be maximized. However, due to the shape of the Choquet integral (not continuous and not differentiable function), there does not exist an optimization technique that exactly solves this problem. Thus, heuristic techniques are used to optimize the Choquet integral in practice. In this paper, we propose the use of a modified genetic algorithm as the optimization technique. Several types of modifications are tested and their performance recorded with respect to behavior

---

* Current address: Family Medicine & Biomedical Sciences, Paul L. Foster School of Medicine, Texas Tech University Health Sciences Center, 9849 Kenworthy Street, El Paso, Texas 79924, USA.

M. Ceberio and V. Kreinovich (eds.), *Constraint Programming and Decision Making*,   97
Studies in Computational Intelligence 539,
DOI: 10.1007/978-3-319-04280-0_13, © Springer International Publishing Switzerland 2014

of the Choquet integral. Before describing the proposed modifications, we review basics of multi-criteria decision making (Section 2) and genetic algorithms (Section 3) as well as previous attempts to modify genetic algorithm to address some of its drawbacks (Subsection 3.1). In Section 4, we propose new modification to the genetic algorithm, and present the experimental setting (Section 5) and the results of the experiments (Section 6), as well as the recommendation when to use each type of modifications (Section 7).

2   Multi-criteria Decision Making

Multi-criteria decision making seeks an optimal solution among a (finite) set of alternatives that are characterized by several attributes, i.e., criteria. Each individual criterion can take a set of values, which could be ordered based on the preference of a decision-maker. For example, if an individual wants to buy a car, this individual will most probably consider the price, speed, miles per gallon, safety rating, color, and other characteristics of cars. A rational car buyer would prefer a low price, a possibility to drive more miles per gallon, and a high safety rating. Thus, he/she would like to select the alternative (i.e., a car) that posses the best value of each criterion (i.e., the cheapest car, the car with the highest safety rating, etc.). However, in reality this is usually not possible. The cheapest car does not have the highest safety rating. Thus, a decision-maker needs to select an alternative that does not posses all the "perfect" characteristics. The natural way to accomplish this goal is to combine the preferences over individual attributes into a global preference over alternatives.

The criteria that a decision maker considers could be qualitative, such as color of a car, or quantitative. Quantitative criteria could take continuous values in some range, such as the price of a car, or could take discrete values, such as the number of stars given for safety rating. Typically, it is possible to map the values of each criterion onto a common scale, which is usually the range $[0, 1]$, where 0 represents the lowest preference and 1 represents the highest preference.

The next task is to combine the values of individual preferences into a global preference. Numerous methods exists to combine partial preferences into a global preference. Simple methods, such as maximum and minimum, and additive methods, such as weighted sum, are usually not good aggregation operators. While the former two methods rely on ordering alternatives based on only one criterion, weighted sum considers all the criteria, but ignores dependencies among the criteria. However, in reality, many criteria are not independent. For example, the price of a car usually increases as the safety rating of the car increases, and the safety rating decreases as the maximum speed increases.

Thus, we need to use non-additive approaches to take into consideration dependencies. Non-additive (or fuzzy) measure theory is an extension of traditional measure theory, which allows us to define integrals that take into consideration dependencies among criteria. The Choquet integral is one of these integrals, but its complexity makes it infeasible in many situations. However, the Choquet integral with respect to a 2-additive measure reduces the computational complexity and still takes into account dependencies among criteria.

Definition 1. *The* Choquet integral *with respect to a 2-additive measure* μ *is given by* [10]:

$$(C)\int_I fd\mu = \sum_{I_{ij}>0}(f(i)\wedge f(j))I_{ij} + \sum_{I_{ij}<0}(f(i)\vee f(j))|I_{ij}|+$$

$$\sum_{i=1}^{n}f(i)\left(I_i - \frac{1}{2}\sum_{j\neq i}|I_{ij}|\right) \qquad (1)$$

where I_i *is the importance of the criterion* i, *usually given by the Shapley value* [20], *and* I_{ij} *is the interaction between the criteria* i *and* j, *usually given by interaction index of order 2* [7, 11].

While Shapley value is given in the interval $[0, 1]$ with 0 representing no importance and 1 being the highest importance, interaction index can take values in the interval $[-1, 1]$, where

- $I(i, j) > 0$ if the criteria i and j are complementary;
- $I(i, j) < 0$ if the criteria i and j are redundant;
- $I(i, j) = 0$ if the criteria i and j are independent.

The function $f(i)$ is the value of the criterion i mapped on the interval $[0, 1]$. However, this mapping might depend on some variables, which are to be optimized.

The goal of multi-criteria decision making is to select an alternative for which the Choquet integral attains the highest value. However, the Choquet integral is often not a differentiable function since it requires calculating maximum or minimum of two functions, and therefore many standard optimization techniques that use differentiation are not applicable. Thus, heuristic techniques are used to maximize the Choquet integral. Genetic algorithm is one of these techniques.

Moreover, the solution to an optimization problem is often limited by some constraints. The most common constraint is that all variables should be non-negative, but many other constraints could be imposed. Thus, the optimization function used in a genetic algorithm should be modified to take into account the constraints.

3   Basics of a Genetic Algorithm

Before explaining possible modifications to a genetic algorithm to suit better the optimization of the Choquet integral, we first review the basics of a classical genetic algorithm.

Genetic algorithms (GAs) are heuristics used as an optimization technique or a machine learning technique. Their performance is based on natural "survival of the fittest" and biological inheritance in organisms. Therefore, they imitate the processes of reproduction through the selection of the fittest individuals, crossover, and mutation. The genetic algorithms were first introduced by John

H. Holland [13] in early seventies, and since then they have found applications in different areas including bioinformatics, chemistry, finance, scheduling, design, etc. [5, 9].

A genetic algorithm simulates a biological evolution through generations. Therefore, it starts with generating an initial population of individuals. Each individual is characterized by several values that are encoded in genes. The fittest individuals are selected for crossover. The fitness of an individual in a genetic algorithm is defined by an objective function that needs to be optimized. The selection of individuals for crossover is usually carried out through the roulette wheel method, which assigns to each individual probability of being selected by the individual's fitness relative to the fitness of the entire population.

Crossover of selected individuals allows for exchange of genetic material. The simplest and the most traditional method for its simulation in a GA is a one-point crossover technique in which a random point among genes is selected. The first offspring is created by copying the genes up to the crossover point from the first parent and the remaining genes from the second parent, while the reverse copy of the parental material is used to create the second offspring. Two offsprings are expected to have higher fitness than their parents which are discarded. However, that is not always the case, so elitist strategy is often implemented to copy at least the current best individual to the next generation. The non-overlapping generations, which requires all parents to be replaced, in combination with elitist strategy is used in experiments performed in this paper.

Further, usually with a very small mutation rate, a mutation may occur, which allows for genetic material inherited from parents to be changed. In terms of genetic algorithms, mutations allow for unexploited areas of search space to be visited. One-point mutation is the most commonly used mutation technique in a GA, where a gene is randomly selected and assigned a random value from the range of values that gene can take.

Generations evolve indefinitely. However, for a genetic algorithm to be useful in practice, either a convergence criterion should be achieved (all individuals have same fitness) or a predefined number of generations (i.e., iterations) have been reached. In our case, we will not test for the convergence criterion but rather only the highest fitness achieved by an individual in a reasonable number of iterations.

3.1   Modification of Genetic Algorithms

Genetic algorithm represents a relatively quick method to finding an optimum and in most cases the results of a GA performance are better than the results obtained by other optimization techniques. However, GA is not without drawbacks. The main drawbacks include that the solution might be only local and not global solution, the solution is not exact, and the speed of the convergence. Since the earliest design of genetic algorithm, several methods have been suggested to overcome these problems including increasing the size of population [6], using different crossover operators that allow individuals to exploit new regions [4], increasing the mutation rate [6], modifying the fitness assignments through fitness scaling and sharing [15], using reserve selection [2], tracking changing environments [3, 12], and restarting [1].

Despite all the proposed modifications, the most common and the most serious drawback of a genetic algorithm still remains its premature convergence, which leads to trapping the solution in a local optimum rather than a global one [5]. One of the main reasons of premature convergence is that the part of the search space containing the global optimum value is not exploited. The usual attempts to reach unexploited parts of the search space occur through mutations. However, with a small probability of mutation occurring, it is not likely that all parts of the search space could be reached. Thus, it is of crucial importance to initialize and update the population in such a way that every part of search area could be exploited. Several ideas have been developed on how to generate the initial population including splitting the entire search space into subspaces of equal sizes known as latin hypercube sampling [17], quasi random sampling [18], dividing search space into subspaces based on population divergence [21], nonaligned systematic sampling [19], simple sequential inhibition [8], and including some particular individuals [14].

Four different techniques have been tested on a set of general functions [16] and results were reported on the coverage of the search space, genetic diversity of individuals in population, and speed of convergence. The four techniques included pseudo-random sampling, Niederreiter generator, simple sequential inhibition process, and nonaligned systematic sampling. Pseudo-random sampling is the most commonly used technique. It relies on pseudo-random generator to generate diverse population. Niederreiter generators represent a quasi sampling method, whose goal is to produce points that maximally avoid each other. Simple sequential inhibition process allows a new individual to enter the population only if its distance from each individual already in the population is greater than some predefined value Δ. Nonaligned systematic sampling divides search space in hypercubes and generates an individual in each subspace.

The results show that the pseudo-random sampling is fast and produces a great genetic diversity, but it usually does not cover the entire search space and does not allow population to exploit the entire search space. Nonaligned systematic sampling does not produce expected genetic diversity, but is able to exploit the entire search space. Niederreiter generators outperformed pseudo-random sampling in terms of search space coverage, but not in terms of genetic diversity, while this quasi sampling method outperformed nonaligned systematic search in genetic diversity but not in coverage of search space. Finally, simple sequential inhibition process performed very well in both genetic diversity and search space coverage criteria, but is a very slow algorithm.

Different modification of genetic algorithms have different impacts on the final outcome of an optimization. These impacts are greatly dependent on the type of the function optimized. In this paper, we focus on modifying the basic genetic algorithm in order to maximize the function represented by the Choquet integral with respect to a 2-additive measure.

4   Modified Genetic Algorithm to Suit Applications in Decision Making

We test two approaches for generating the initial population that will allow each point of search space to be reached in short time: quasi sampling and adding special individuals. We also compare the results of these functions to the result of a classical genetic algorithm whose initial population is generated pseudo-randomly.

The first approach generates "extreme" individuals. As the name suggests, these individuals contain the limiting (extreme) values at each gene. If for each gene i, we represent the values it can take by the interval $[a_i, b_i]$, then two main extreme n-gene individuals would be $(a_1, a_2, \ldots, a_n)$ and $(b_1, b_2, \ldots, b_n)$. Of course, we can create more extreme individuals by selecting either a_i or b_i for each gene i. If considered in two-dimensions (i.e., an individual contains only two genes), two extreme individuals would correspond to the corners of the quadrilateral that are diagonally positioned form each other. In two-dimensional space, only two more extreme individuals could be created, and they would correspond to the other two corners of the quadrilateral. In three-dimensional search space, total of eight extreme individuals could be created corresponding to eight corners of polyhedron. In n-dimensional search space, the maximum number of extreme individuals is 2^n. In the next section, we explore in which cases and how well the existence of two or more extreme individuals improves the performance of genetic algorithm. For that reason we have implemented five different algorithms.

The first algorithm 2EXT creates only two main extreme individuals, which are treated as all the other individuals in the population. The next two algorithms consider these two individuals as special individuals in the population. For each extreme individual, an individual from the population is selected and the crossover is performed creating total of four new individuals that replace randomly selected individuals in the population. This process is repeated ten times in 10ITER algorithm and twenty times in 20ITER algorithm. After these initial 10 or 20 crossovers are performed, the algorithm follows the steps of a classical genetic algorithm:

```
create two extra individuals;
randomly initialize the remaining part of population;
for(i=0;i<10;i++) //for 20ITEM, the limits is 20 instead of 10
{
    select an individual x from population;
    crossover x with one extreme individual;
        //it yields two children
    crossover x with the other extreme individual;
        //two more children
    replace four randomly chosen individuals in the population;
        //do not replace extreme individuals;
}
start the classical genetic algorithm;
```

The fourth proposed algorithm, 8EXT, utilizes eight extreme individuals. The first two extremes are created in the same manner as in the 2EXT case, and the remaining six extremes are generated by randomly picking either the lower or the upper bound of the range of values for each gene. Finally, 2^nEXT algorithm generates all possible extreme individuals. However, this algorithm is only applicable to situations with small number of genes since creating all individuals in 20- or 30-dimensional space would create population whose size is enormous, thus creating just one generation in this situation would take extremely large time and space.

The idea of having extreme individuals is that when they are involved in crossover, they allow for generation of offsprings that are in areas possibly not covered by any other individual. Thus, the main reason to start with extreme individuals is to early generate population that is spread over the entire search space and covers (almost) every existing "corner". In general, for extreme individuals to have a significant effect on the performance of a genetic algorithm, they need to participate in numerous crossover processes; thus, we created the algorithms 10ITER and 20ITER, which perform several crossovers between each extreme individual and any individual in the population before the extreme individuals are considered in the same manner as all the other individuals. We test how much improvement in the performance of genetic algorithm is accomplished when different number of initial crossovers with extreme individuals is performed.

The second approach to generating initial population that can reach any point in the search space is by splitting the entire search space in subspaces and requiring that at least one individual is randomly generated in each subspace. If we make the assumption that all dimensions have the same ranges of values that genes can take, in two-dimensional space, the subspaces would be small squares covering the entire search space; in three dimensions, small cubes that cover the entire search space, etc. The subspaces are equally spaced, so having an individual in each subspace creates a population that occupies every part of the search space. However, this algorithm becomes infeasible when the number of genes is large as we need to create large number of subspaces (or a moderate size) to cover the entire search space. If we create small number of subspaces that are large, we are not guaranteeing that the population will be spread well. The algorithm SPLIT that was used for the comparison to other algorithms used four subintervals in each dimension when five-genes individuals were tested, and eight subintervals when three-genes individuals were tested. In the next section, we also explore the impact of the number of subspaces on the performance of the algorithm.

5 Experiments

Two different functions with multiple coefficients were tested. The first function takes the form that resembles what is happening within the Choquet integral with respect to a 2-additive measure. We denote by x[i] the value that i^{th}

gene takes, which would correspond to the utility value of the criterion i. For simplicity, calculation of maximum and minimum of two value is alternated. After finding maximum or minimum value of a pair of utility values, this value is multiplied by a randomly generated number in the interval $[-1, 1]$, which would correspond to the interaction index of order two. The function is optimized with respect to variables x[i]. The pseudocode of the function follows:

```
f = 0;
for(i=0;i<(numberOfGenes-1);i++)
    for(j=(i+1);j<numberOfGenes;j++)
    {
        if ((i mod 2) == 0)
            f=f+a*min(x[i],x[j]);
        else
            f=f+b*max(x[i],x[j]);
    }
return f;
```

The average maximum obtained in 20 runs of this algorithm are reported in table 1.

The second function is even closer representation of the Choquet integral with respect to a 2-additive measure, where the optimum value has to be reached while satisfying numerous (soft) constraint. This function also involves calculating max or min of pairs of points, but in this case the points compared are not directly values of genes but rather values of functions (called parameters) whose inputs are values of particular genes. In addition to added complexity in terms of elements for comparison, this function also simulates solving problem with (soft) constraints by assigning penalties when constraints are not satisfied. Two constraints imposed in this problem are:

1. Sum of values of all genes in an individual can not be greater than 1.
2. Each gene should have a positive value.

There is a high penalty imposed if the first constraint is not satisfied, while a lower penalty is imposed for each gene not satisfying the second constraint. Keeping the same notation as in the pseudocode of the previous function, and denoting by sums[i] the ith parameter for comparison, we present the pseudocode of the (optimizing) function used for testing purposes:

```
for(i=0;i<numberOfGenes;i++)
    sums[i]=0; //initialize functions
for(i=0;i<numberOfGenes;i++) //calculate parameters
    for(j=0;j<numberOfGenes;j++)
        if (((i+j) mod 2)==0) sums[i]=sums[i]+x[j];
        else sums[i]=sums[i]-x[j];
f = 0;
for(i=0;i<(numberOfGenes-1);i++)
```

Optimization of the Choquet Integral   105

```
//calculate optimizing function
    for(j=(i+1);j<numberOfGenes;j++)
        if (((i+j) mod 2)==0)
            f=f+(min(sums[i],sums[j]))*sums[i];
        else
            f=f-(max(sums[i],sums[j]))*sums[j];
if (sum>1) f=f-1000000; //first constraint
for(i=0;i<numberOfGenes;i++)
    if (x[i]<=0) f=f-1; //second constraint
```

The average maximum obtained in 20 runs of this algorithm are reported in Table 2.

Both functions, are tested using 3, 5, 10, 20, and 30 genes in an individual, where each gene could take values in the range $[0, 1]$.

6   Results

In this section, we tabulate the results obtained from the experiments and give an explanation for the best performing variations of the genetic algorithm. Dashes are placed in the entries in the table for which the experiments were not performed either due to large initial population that would result from the particular method (in the case of 2^nEXT and SPLIT algorithms) or due to achieving the global optimum value by the initial population since the global optimum is attained by one of the "extreme" individuals (in the case of 2EXT, 10ITER, 20ITER, and 8EXT algorithms). For the later reason, some algorithms are not included in particular tables.

Table 1. Maximum obtained for the first function

method	GA	2EXT	10ITER	20ITER	8EXT	SPLIT
3 genes	1.99994	—	—	—	—	**2.00000**
5 genes	12.9981	12.9995	12.9995	12.9994	**12.9999**	12.9990
10 genes	129.896	129.980	129.989	129.993	**129.999**	—
20 genes	1182.97	1183.20	**1184.30**	1184.10	**1184.30**	—
30 genes	4148.36	4151.45	4155.67	4158.49	**4163.35**	—

All the modified algorithms outperformed the classical genetic algorithm when tested on the first function. As expected, the best performance was achieved by the 8EXT algorithm as the global optimum seems to lay along the boundary of the given interval values for genes. Also the SPLIT algorithm performed exceptionally well in the application with three genes. However, all the other algorithms achieved very close results, so we can not make any conclusions about the algorithms based on the results from testing the first function. However, the second function gives much more insides in the performance of the presented algorithms.

Table 2. Maximum obtained for the second function

method	GA	2EXT	10ITER	20ITER	8EXT	2^nEXT	SPLIT
3 genes	**0.99988**	**0.99988**	**0.99988**	**0.99988**	**0.99988**	**0.99988**	0.99987
5 genes	3.99731	3.99854	**3.99902**	**3.99902**	3.99805	3.99902	3.99896
10 genes	-999376	21.5746	**23.4975**	23.4567	**23.4130**	20.4190	—
20 genes	-990082	60.6868	62.7694	**82.3662**	**85.9714**	—	—
30 genes	-950297	155.596	**196.492**	**196.180**	192.350	—	—

When testing algorithms using the second function, we can see that all algorithms except the SPLIT algorithm when only three genes are used outperformed the classical GA. The 20ITER algorithm is constantly among the top two performers, and even when it yields the second best answer, the reached optimum is not far from the best optimum obtained by any of the algorithms.

Also, looking at the results of the classical GA, when the number of genes increases, it is easy for this algorithm to be trapped in a local optimum and not be able to overcome the penalties set for not satisfying constraints. We can see this fact from the negative values in the Table 2, which resulted from the inability of the algorithm to satisfy the first constraint.

6.1 Further Examination of the SPLIT Algorithm

The SPLIT algorithm is feasible only with small (i.e., ≤ 6) number of genes in an individual. However, its performance is very good in these cases, so we spent some time trying to conclude how many subintervals along each dimension would result in the best performance of the algorithm. We tested three-genes individuals on the two functions, by splitting each dimension into 4, 5, and 8 subintervals. However, the obtained results (summarized in Table 3) were not conclusive.

In both functions, it seems that the best results were obtained when eight subintervals were used. However, when we compare the results from using four intervals to results when five intervals are used, the first and function behaved better when the interval was split into four subintervals, while the second function performed better when five intervals were used. We would like to conclude that using eight subintervals shows the best performance. However, we can not make this conclusion as there is not enough evidence that higher the number of subintervals leads to the better performance of the algorithm. Moreover, when considering individuals that contain five or six genes, splitting intervals in eight subintervals would result in generating a population of over thousand individuals, which might not be feasible.

6.2 Statistical Analysis

We performed t-test to determine the significance of the results obtained in the modified algorithms compared to the classical genetic algorithm. The results summarized in Table 4 show the p-values obtained by running one-tailed pairwise

Table 3. Maximum obtained by different versions of the SPLIT algorithm

function	max f_1	max f_2
4 subintervals	**2.2400**	9.07017
5 subintervals	2.2396	**9.27518**
8 subintervals	**2.2400**	**9.28987**

Table 4. p-values obtained in one-tailed pairwise t-test between denoted algorithms and classical GA

2EXT	10ITER	20ITER	8EXT	2^nEXT	SPLIT
0.0403	0.0403	0.0403	0.0403	0.2113	0.09803

t-test between all the results obtained in denoted algorithms against the classical genetic algorithm.

Based on the obtained p-values, we can conclude at 95% confidence level that proposed modifications in the first four algorithms (2EXT, 10ITER, 20ITER, and 8EXT) outperform the classical genetic algorithm. The 2^nEXT algorithm gives us only 79% confidence level that obtained results are better than in the classical GA, but this fact might be due to a low number of points for comparison. Recall that we can run 2^nEXT algorithm only when the dimension n is relatively small, thus we had no values obtained for $n > 5$. Moreover, we did not run this algorithm for the first function since the results of the first function lie very close to (or on) the boundaries. Finally, the SPLIT algorithm is also applicable only in situations with a small number of n, but we were able to run this algorithm on both functions. We can conclude at 90% confidence level that, when $n \leq 5$, this algorithm outperforms the classical genetic algorithm.

7   Conclusion

The SPLIT algorithm showed good performs when three or less genes were used to represent an individual. A little bit different setting allows for spreading individuals along (or closer to) the edges of the n-polytop representing the search space rather than everywhere close to boundaries and inside the polytop. This setting also allows for all ares of search space to be exploited since crossover of two individuals could lead to production of an individual in the interior of the search space. This is part of the future work.

The 8EXT algorithm performed the best when the genes' values of the optimal solution lay along the boundaries of the intervals of possible gene values. However, when not-knowing what to expect from the optimizing function (as it is usually the case when modeling real life situations), the 10ITER and 20ITER algorithms represent the best bet. They had outstanding performance in any situation. The 10ITER is better suited for smaller number of genes, while 20ITER is needed to perform optimization when higher number of genes is present in an individual.

References

1. Beligiannis, G., Tsirogiannis, G.A., Pintelas, P.E.: Restarting: a technique to improve classic genetic algorithms' performance. In: Proceedings of World Academy of Science, Engineering and Technology, vol. 1, pp. 144–147 (2005)
2. Chen, Y., Hu, J., Hirasawa, K., Yu, S.: Performance tuning of genetic algorithms with reserve selection. In: IEEE Congress on Evolutionary Computation, pp. 2202–2209 (2007)
3. Cobb, H.G.: Genetics algorithms for tracking changing environments. In: Proceedings of the Fifth International Conference on Genetic Algorithms, pp. 523–530. Morgan Kaufmann (1993)
4. Combarro, E.F., Miranda, P.: Identifying of fuzzy measures from sample data with genetic algorithms. Computational Operational Reseach 33(10), 3046–3066 (2006)
5. Davis, L.: Handbook of Genetic Algorithms. Van Nostrand Reinhold, New York (1991)
6. De Jong, K.A.: An Analysis of the behavior of a class of genetic adaptive systems. PhD thesis (1975)
7. Denneberg, D., Grabisch, M.: Shapley value and interaction index. Mathematics of interaction index (1996)
8. Diggle, P.J.: Statistical analysis of spatial point patterns. Academic Press, London (1983)
9. Goldberg, D.E.: Genetic Algorithms in Search, Optimization, and Machine Learning. Addison-Wesley Professional (1980)
10. Grabisch, M.: The interaction and Mobius representation of fuzzy measures on finite spaces, k-additive measures: a survey. In: Grabisch, M., Murofushi, T., Sugeno, M. (eds.) Fuzzy Measures and Integrals: Theory and Applications. Physica Verlag (2000)
11. Grabisch, M., Roubens, M.: Application of the Choquet integral in multicriteria decision making. In: Garbisch, M., Murofushi, T., Sugeno, M. (eds.) Fuzzy Measures and Integrals: Theory and Applications. Physica Verlag (2000)
12. Grefenstette, J.J.: Genetic algorithms for changing environments. In: Parallel Problem Solving from Nature 2, pp. 137–144. Elsevier (1992)
13. Holland, J.H.: Adaption in natural artificial systems. University of Michigan Press, Ann Arbor (1975)
14. Karci, A.: Novelty in the generation of initial population for genetic algorithms. In: Negoita, M.G., Howlett, R.J., Jain, L.C. (eds.) KES 2004. LNCS (LNAI), vol. 3214, pp. 268–275. Springer, Heidelberg (2004)
15. Kreinovich, V., Quintana, C., Fuentes, O.: Genetic algorithms: what fitness scaling is optimal. Cybernetics and Systems: an International Journal 24, 9–26 (1993)
16. Maaranen, H., Miettinen, K., Penttinen, A.: On initial populations of a genetic algorithm for continuous optimization problems. Journal of Global Optimization 37(3), 405–436 (2007)
17. McKay, M.D., Beckman, R.J., Conover, W.J.: A comparison of three methods for selecting values of input variables in the analysis of output from a computer code. Technometrics 21(2), 239–245 (1979)

18. Niederreither, H.: Random Number Generation and Quasi-Monte Carlo Methods. SIAM, Philadelphia (1992)
19. Ripley, B.D.: Spatial statistics. John Wiley & Sons, New York (1981)
20. Shapley, L.S.: A value for n-person games. In: Kuhn, H.W., Tucker, A.W. (eds.) Contributions to the Theory of Games, vol. 2, pp. 307–317. Princeton University Press (1953)
21. Tsutsui, S., Fujimoto, Y., Ghosh, A.: Forking genetic algorithms: GAs with search space division schemes. Evolutionary Computation 5(1), 61–80 (1997)

Scalable, Portable, Verifiable Kronecker Products on Multi-scale Computers

Lenore Mullin[1] and James Raynolds[2]

[1] College of Computing and Information (CCI), University at Albany
State University of New York, Albany, NY
lenore@albany.edu
[2] Drinker Biddle & Reath, L.L.P., Washington. D.C., USA
james.raynolds@dbr.com

Abstract. Understanding the layout of data and the accessing of that data is paramount to the optimal performance of an algorithm on one or many processors. This paper addresses the need for efficient tools to implement and carry out tensor based computations for scientific and engineering applications. In particular, we focus on certain ubiquitous operations such as outer products of arbitrary multi-dimensional arrays and matrix Kronecker products. We advocate an algebraic methodology based on A Mathematics of Arrays (MoA) and the ψ-Calculus, in which, any array based computer language (such as MATLAB) would be augmented to achieve optimal performance for the computation of multiple outer products. In this approach, an Operational Normal Form (ONF), which specifies the most efficient implementation in terms of *starts*, *stops*, and *strides* is mathematically derived given specific details of the processor/memory hierarchy. The vision of this research is the creation of a system in which the application scientist or engineer can use a functional subset of his/her favorite language and, in so doing, have the ability to generate code with high efficiency and compiler-like optimizations.

1 Introduction

High dimensional computational modeling is ubiquitous across the sciences and engineering. Due to its support of matrix (and more recently tensor) operations [14], MATLAB is used as a rapid prototyping language. The interactive mode of work is appealing to experimenters because it allows them to get initial results without a lengthy edit, compile, link, execute cycle.

Currently there is considerable demand for efficient methods to handle tensor based computing problems. According to van Loan: "The scientific and engineering communities are awash in a sea of high-dimensional, multi-indexed data sets. Tensor methods are required to expose the underlying patterns...[p]ortability, reusability, reliability, correctness, and modularity need to be reconciled with the computational scientist's need for efficiency, especially on massively parallel multi-core architectures." [31, 32].

M. Ceberio and V. Kreinovich (eds.), *Constraint Programming and Decision Making*,
Studies in Computational Intelligence 539,
DOI: 10.1007/978-3-319-04280-0_14, © Springer International Publishing Switzerland 2014

At a recent NSF sponsored workshop held in Edinburgh, Scotland, co-funded by Edinburgh's Center for Numerical Algorithms and Intelligent Software (NAIS) [25], three levels of interaction were identified, which, in somewhat simplistic terms, can be summarized as the triple A's:

$$\text{Application} \leftrightarrow \text{Algorithm} \leftrightarrow \text{Architecture}.$$

In the present work we discuss ways to augment MATLAB, *or any array/tensor language,* to achieve optimal performance for multiple Kronecker Products. Multiple Kronecker Products are widely used in many application areas. In high-energy physics, for example, we find tensor based data sets arising from outer products such as: *space-time* $\otimes$ *color* $\otimes$ *spin*. High-energy physics computations consume the resources of today's largest most advanced multi-core computers and, as such, represent a good testing ground for the tensor based research discussed herein.

An important realization is the need to identify *core algorithms* so as to:

- optimize multiple instantiations,
- scale the designs,
- map to complex processor memory hierarchies, and,
- verify correctness and thus guarantee computational reproducibility.

Widespread Use of MATLAB. MATLAB is a matrix-based high-level interactive language that is used in many disciplines in industry, academia, and government as a rapid prototyping language (e.g. it is used in electrical engineering, image processing, aerospace engineering, physics, materials science, bioinformatics, Simulink and FPGA design, etc.). For image processing applications the ability to handle 2D arrays of data and operate on them as *first class objects* (i.e. *monolithic entities with whole-array operations*), greatly improves productivity. Moreover, when arrays/tensors are first class data objects, the structure, shape and dimensionality, of the tensor can aid in compilation.

The development of toolboxes and specialized libraries has also provided the end user with tools for enhanced productivity. Recent awareness to n-dimensional arrays, (i.e. tensors), as well as new insights into optimal implementations [31], has led scientists to develop a Tensor Toolbox for MATLAB [3–7, 14], thus providing a familiar environment that supports tensors and has motivated subsequent algorithm research.

The Limitation with Respect to Compiled Performance. MATLAB, like any interpretive language, is limited in the speed at which it can run programs when compared with compiled machine code. Earlier versions of MATLAB provided the facility to translate to C, but trying to *add on* a compiler in this way has typically met with limited success. The C code produced could only be run when linked with MATLAB proprietary libraries which were available only under license, limiting its use as a general code dissemination tool. Translation to C also threw away information about the parallelism that was present in the original *whole-array*, matrix-valued MATLAB expressions. As a result, when the C

code was compiled it was hard for the C compiler to deduce when it was safe
to use parallel hardware which might be present on the target processor. Consequently, parallel MATLAB [13] was developed to give the user a tool to parallelize
MATLAB programs using MPI.

Our research advocates *an environment that stays at the highest level
of mathematical interface and gives the user the highest level of
performance, correctness, and scalability, even to multi-core hybrid
systems.*

When compiler support was implemented in APL, MATLAB's predecessor,
the main limitation was that the anomalies in the language [10] made it difficult
to optimize and impossible to verify correctness of programs. MATLAB has data
structures (matrices), and anomalies similar to those of APL, and consequently,
solutions similar to those found for APL could be applied in the context of MAT-
LAB (*or any other array-supporting language such as Fortran95, Phython, etc.*).
Thus, it may be advantageous to identify a functional array sublanguage [29] that
can act as an intermediate representation in a compiler pipeline. This representation would retain information about parallelism while allowing high-level optimizations to be made that capitalize on locality and successful data prefetching.
As an interpretive language, MATLAB also makes less efficient use of cache than
is possible with a compiler which, ideally, minimizes the formation of temporary
array values [28].

2   Background and Initial Goals

Relevant Techniques and Background. In order to address such questions in
the compiling of array languages, Mullin developed a Mathematics of Arrays
(MoA) and the ψ-Calculus (an indexing calculus based on shapes) [17]. This
research was motivated by the need to remove the anomalies in Iverson's array
algebra [1] and to put closure on Abram's ideas of defining all array operations
using shapes [2]. Such ideas were inspired by the work of Alan Perlis [30] and
Klaus Berkling [8]. Moreover, MoA and ψ-Calculus, can be used to abstract
complex processor memory layouts for tensor/array expressions, reduce these
expressions first to a semantic normal form (Denotational Normal Form-DNF)
then to an operational normal form (ONF). The ONF describes how to *build* the
code using *start, stop, stride,* and *count*: a universal machine abstraction [9, 18,
21, 24, 26, 27].

First Steps. The two workshops mentioned above: at Edinburgh [12], and at
NSF [31] illustrated ways in which researchers are attempting to meet the chal-
lenges posed by large scale tensor based computing. The Edinburgh workshop
report proposed a ten year challenge to design, verify, scale, and port a few
important algorithms from one or two highly used languages to one or two architectures using an abstract machine.

Our research advocates choosing heavily used production languages and data
structures. Our hopes are to engage the language designers in our efforts and to
create a team of collaborators that can cooperatively solve this problem.

3   The ψ-Calculus

Mechanization. MoA and the ψ-Calculus provide a way to *compose* array operations to minimize intermediate/temporary arrays. The ψ-Calculus provides a *mechanical* way to compose indexing operations using shapes. Consequently, we can use this theory to *hand derive* designs until tools, compilers, libraries, and languages are developed. Although initial attempts have been made to *mechanize* ψ-reduction [11, 15, 19, 20, 22–24], no tool or language mechanizes the designs developed so far by hand [21].

Core Algorithms

The Kronecker Product. In discussing the importance of the Kronecker Product, Charles Van Loan quips that it is "the product of the times" and argues that this product is ubiquitous in science and engineering computation [33]. The need for efficient computation of multiple products pervades science, in particular, quantum computing [16] and high energy physics [25]. Since there are numerous descendants [33] of the Kronecker Product: The Left Kronecker Product, The Hadamard Product, The Tracy-Singh Product, The Khatri-Rao Product and the Generalize Kronecker Product, a generalized design and verification is worth pursing.

Outer Product. A key notion of the present work is how the MoA *outer product* can be formulated as the Kronecker product, a special case of the *tensor product*. We will show that the use of the MoA outer product is superior to the traditional approach when one is concerned with efficient implementations of multiple Kronecker products. The MoA outer product is a general operation on two arrays of any shape or dimension and applies any scalar operation, not just multiplication ($*$), to these two arrays (i.e. $+$, $-$, $/$, etc. are valid operations). Using the ψ-Calculus (a calculus of indexing), we can compose multiple outer products such that the creation of intermediate, temporary arrays is minimized or *eliminated*. The first step in this procedure results in a semantic normal form, the Denotational Normal Form (DNF) which specifies the algorithm in terms of Cartesian array indices. Then the DNF to is translated into the Operational Normal Form (ONF) which specifies how to build the code in terms of *starts*, *stops*, and *strides*: an ideal machine abstraction. The ONF is obtained from the DNF by utilizing specific information regarding the layout of the machine in terms of complex processor memory hierarchies.

This is possible by *abstractly* lifting the dimension of an expression to include all aspects of a processor, memory, communication topology. In this work we focus on the Kronecker product between two matrices of arbitrary size resulting in a block matrix. Let's begin with an example where:

$$A = \begin{bmatrix} 1 & 2 \\ 3 & 4 \end{bmatrix}, \tag{1}$$

and

$$B = \begin{bmatrix} 5 & 6 & 7 & 8 \\ 9 & 10 & 11 & 12 \\ 13 & 14 & 15 & 16 \end{bmatrix}. \tag{2}$$

The operation $A \otimes B$, is called the Kronecker product and is defined as:

$$\begin{bmatrix} 1 & 2 \\ 3 & 4 \end{bmatrix} \otimes \begin{bmatrix} 5 & 6 & 7 & 8 \\ 9 & 10 & 11 & 12 \\ 13 & 14 & 15 & 16 \end{bmatrix} =$$

$$\begin{bmatrix} 1 \times 5 & 1 \times 6 & 1 \times 7 & 1 \times 8 & 2 \times 5 & 2 \times 6 & 2 \times 7 & 2 \times 8 \\ 1 \times 9 & 1 \times 10 & 1 \times 11 & 1 \times 12 & 2 \times 9 & 2 \times 10 & 2 \times 11 & 2 \times 12 \\ 1 \times 13 & 1 \times 14 & 1 \times 15 & 1 \times 16 & 2 \times 13 & 2 \times 14 & 2 \times 15 & 2 \times 16 \\ 3 \times 5 & 3 \times 6 & 3 \times 7 & 3 \times 8 & 4 \times 5 & 4 \times 6 & 4 \times 7 & 4 \times 8 \\ 3 \times 9 & 3 \times 10 & 3 \times 11 & 3 \times 12 & 4 \times 9 & 4 \times 10 & 4 \times 11 & 4 \times 12 \\ 3 \times 13 & 3 \times 14 & 3 \times 15 & 3 \times 16 & 4 \times 13 & 4 \times 14 & 4 \times 15 & 4 \times 16 \end{bmatrix}. \tag{3}$$

Note implicitly in the operation above, that the 4 multiplications applied to B have a substructure within the resultant array. That is, **each** component of A is multiplied with **all** of B creating 4, 3×4 arrays. The result is stored in a matrix, C, by relating the indices of A, i,j, with the indices of B, k,l, and encoding them into row, column coordinates. Classically, i,k is correlated to a row of C, and j,l is correlated to a column. In other words we write $C_{M,N} = A_{i,j} * B_{k,l}$ where $M = \{i, k\}$, and $N = \{j, l\}$, are composite indices advancing in row-major ordering (i.e. $\{i, j\} = \{0, 0\}, \{0, 1\}, \{0, 2\}, \cdots$).

An important goal of this paper is to describe how **shapes** are integral to array/tensor operations. By definition, the **shape** of an array is a vector whose elements equal the length of each corresponding dimension of the array. Using shapes, we will relate operations in A Mathematics of Arrays (MoA) to tensor algebra and we will show how these shapes and the ψ-Calculus (also sometimes written: Psi-Calculus) can be used to compose multiple Kronecker products and map such operations to complex processor/memory hierarchies. We'll also discuss how easily this can be embodied in an algebraic subset of MATLAB.

Shapes and the ψ Operator. Let's begin by introducing shapes. The shape of A is 2×2, i.e., $\rho A = \langle 2\ 2 \rangle$, the shape of B is 3×4, i.e. $\rho B = \langle 3\ 4 \rangle$ and the shape of $A \otimes B$ is 6×8, i.e. $\rho(A \otimes B) = \langle 6\ 8 \rangle$. In this discussion we have introduced the *shape* operator, ρ, which acts on an array and returns its shape vector.

Now, let's look at the MoA outer product of A and B, denoted by $A\ op_\times B$. The shape of $A\ op_\times B$ is the concatenation of the shapes of A and B, i.e. a 4-dimensional array with shape $2 \times 2 \times 3 \times 4$. That is, $\rho\ (A\ op_\times B) = \langle 2\ 2\ 3\ 4 \rangle$. The resulting array is indexed by a vector $\langle i\ j\ k\ \ell \rangle$ that is ordered in row-major order (i.e. in the order of a nested $\{i\ j\ k\ \ell\}$ loop with ℓ the fastest and i the slowest increasing *partial index*).

The MoA array operation: $A \; op_\times \; B$ is defined by the following result:

$$
\begin{bmatrix} 1 & 2 \\ 3 & 4 \end{bmatrix} op_\times \begin{bmatrix} 5 & 6 & 7 & 8 \\ 9 & 10 & 11 & 12 \\ 13 & 14 & 15 & 16 \end{bmatrix} =
$$

$$
\left[
\begin{array}{l}
\left[\begin{bmatrix} 1\times 5 & 1\times 6 & 1\times 7 & 1\times 8 \\ 1\times 9 & 1\times 10 & 1\times 11 & 1\times 12 \\ 1\times 13 & 1\times 14 & 1\times 15 & 1\times 16 \end{bmatrix} \begin{bmatrix} 2\times 5 & 2\times 6 & 2\times 7 & 2\times 8 \\ 2\times 9 & 2\times 10 & 2\times 11 & 2\times 12 \\ 2\times 13 & 2\times 14 & 2\times 15 & 2\times 16 \end{bmatrix}\right] \\
\left[\begin{bmatrix} 3\times 5 & 3\times 6 & 3\times 7 & 3\times 8 \\ 3\times 9 & 3\times 10 & 3\times 11 & 3\times 12 \\ 3\times 13 & 3\times 14 & 3\times 15 & 3\times 16 \end{bmatrix} \begin{bmatrix} 4\times 5 & 4\times 6 & 4\times 7 & 4\times 8 \\ 4\times 9 & 4\times 10 & 4\times 11 & 4\times 12 \\ 4\times 13 & 4\times 14 & 4\times 15 & 4\times 16 \end{bmatrix}\right]
\end{array}
\right]. \quad (4)
$$

Notice that the layouts in Eqs. 3 and 4 are very similar. What is different is the bracketing. The result of the MoA outer product is *not* a matrix but is rather a multi-dimensional array. In contrast, the result of the Kronecker product *is* a matrix (i.e. a two-dimensional array). The extra brackets reflect the fact that the result of the outer product is a 4-dimensional array whose shape is obtained by concatenating the shapes of the arguments (i.e. $\langle 2\ 2\rangle$ concatenated to $\langle 3\ 4\rangle$ yields $\langle 2\ 2\ 3\ 4\rangle$).

So, do these arrays have the same layout in memory? The answer is no. What is interesting, however, is that when the Kronecker product is computed it is filled in, in a row major ordering relative to the right argument. The layout, either row or column major, would reflect the access patterns needed to optimize these operations across the processor/memory hierarchy. Let's assume row major. Thus flattening (i.e. creating a vector consisting of the elements of the array in row-major order), the difference in layout is as follows. For the Kronecker product we have:

$$
\langle 1\times 5 \ \ 1\times 6 \ \ 1\times 7 \ \ 1\times 8 \ \ 2\times 5 \ \ 2\times 6 \ \ 2\times 7 \ \ 2\times 8 \ \ \ldots\rangle, \quad (5)
$$

whereas for the MoA outer product we have:

$$
\langle 1\times 5 \ \ 1\times 6 \ \ 1\times 7 \ \ 1\times 8 \ \ 1\times 9 \ \ 1\times 10 \ \ 1\times 11 \ \ 1\times 12 \ \ \ldots\rangle. \quad (6)
$$

Let's consider how languages typically implement these operations. Typically, assuming A, B, and C are defined as n by n arrays, the operation:

$$
A \otimes B \otimes C, \quad (7)
$$

would be accomplished by materializing all of $B \otimes C$ as a temporary array, let's call it $TEMP$. Then $A \otimes TEMP$ would be computed. If n is large, this could use an enormous amount of space.

Now, let's look at how MoA and ψ-Calculus would perform the outer product. Then, we'll discuss how we can restructure the MoA outer product to get the Kronecker product and in so doing we'll be able to **compose multiple Kronecker products efficiently and deterministically over complex processor/memory hierarchies.**

Shapes and the Outer Product. Before beginning, we refer the reader to the numerous publications on MoA and the ψ-Calculus, the most foundational is given in Refs. [17, 21]. We thus take liberty to use operations in the algebra and calculus by example. Only when necessary will a definition be given (as in the following).

Definition 1. *Assume* A, B, C, *are* $n \times n$ *arrays, that is, each array has shape:*

$$\rho A = \rho B = \rho C = \langle n\ n \rangle.$$

Assume the existence of the ψ *operator and that it is well defined for n-dimensional arrays. The* ψ *operator takes as left argument an index vector and an array as the right argument and returns the corresponding component of the array. For a full index (i.e. as many components are there are dimensions) a scalar is returned and for a* partial *index, a sub-array is selected. Then,*

$$D \equiv A\ op_\times (B\ op_\times C)$$

is defined when the shape of D is equal to the shape of $A\ op_\times(B\ op_\times C)$. And the shape of $A\ op_\times(B\ op_\times C)$ is equal to the shape of A concatenated to the shape of $(B\ op_\times C)$ which is equivalent to the shape of A concatenated to the shape of B concatenated to the shape of C. i.e.,

$$\rho D = \rho(A\ op_\times(B\ op_\times C)) = \rho A + \!\!+ \rho(B\ op_\times C)$$
$$= \rho A + \!\!+ \rho B + \!\!+ \rho C = \langle n\ n\ n\ n\ n\ n \rangle$$

Then, $\forall\ i_0, j_0, k_0, l_0, m_0, n_0$ *s.t.*
$0 \le i_0 < n;\ 0 \le j_0 < n;\ 0 \le k_0 < n;\ 0 \le l_0 < n;\ 0 \le m_0 < n;\ 0 \le n_0 < n$

$$\langle i_0\ j_0\ k_0\ l_0\ m_0\ n_0 \rangle\ psi\ D$$
$$= \quad ((\langle i_0\ j_0 \rangle \psi\ A) \times ((\langle k_0\ l_0\ m_0\ n_0 \rangle \psi\ (B\ op_\times C)))$$
$$= \quad ((\langle i_0\ j_0 \rangle \psi\ A) \times ((\langle k_0\ l_0 \rangle \psi\ B) \times ((\langle m_0\ n_0 \rangle \psi\ C))$$

It is easy to see that we can compose as little or as much as we like given the bounds of i_0, j_0, k_0, l_0, m_0 and n_0. We'll return to how to build the above composition. We'll also discuss how to include processor memory hierarchies but first we'll discuss how to make the layout of the Kronecker product equivalent to the layout of the MoA outer product.

Permuting the Indices of the MoA Outer Product. In order to discuss permuting the outer product we must first discuss how to permute an array. One way is through a transpose. We are familiar with transposing an matrix, i.e A^T. We know that $A[j; i]$ denotes $A^T[i, j]$. Let's now discuss how to transpose a matrix in MoA and then how to transpose an array in general.

Definition 2. *Given the shape of A is $m \times n$, i.e. $\rho A = \langle m\ n \rangle$, then A^T is defined when the shape of A^T is $n \times m$. That is,*

$$\rho A^T = \langle n\ m \rangle.$$

Then, for all $0 \le i < n$ and $0 \le j < m$

$$\langle i\ j \rangle \psi A^T = \langle j\ i \rangle \psi A$$

Let's now generalize this to any arbitrary array.

Definition 3. *Given the shape of A is $\langle m\ n\ o\ p\ q\ r \rangle$. Then A^T is defined when the shape of A^T is $\langle r\ q\ p\ o\ n\ m \rangle$. Then for all $0 \le i_0 < r; 0 \le j_0 < q; 0 \le k_0 < p; 0 \le l_0 < o; 0 \le m_0 < n; 0 \le n_0 < m;$*

$$\langle i_0\ j_0\ k_0\ l_0\ m_0\ n_0 \rangle \psi A^T = \langle n_0\ m_0\ l_0\ k_0\ j_0\ i_0 \rangle \psi A$$

A question should immediately come to mind. Can the indices permute in ways other than reversing them? The answer is yes, and in fact any permutation consistent with the shape of the array is achieved by simply permuting the elements of the index vector. That is, for an n-dimensional array there are $n!$ permutations. Note that the definitions for **general transpose** and **grade up** presented below are the same definitions that were proposed to the F90 ANSI Standard Committee in 1993 and were subsequently accepted for inclusion in F95.

Definition 4. *The operator **gradeup** is defined for an n-element vector containing positive integers in the range from 0 to $n-1$ in any order (multiple entries of the same integer are allowed). The result is a vector denoting the positions of the lowest to the highest such that when the original vector is indexed by the result of grade up, the original vector is sorted from lowest to highest.*

Example. Given $\vec{a} = \langle 2\ 0\ 1\ 3 \rangle$,

$$gradeup[\vec{a}] = gradeup[\langle 2\ 0\ 1\ 3 \rangle] = \langle 1\ 2\ 0\ 3 \rangle.$$

Thus,

$$\vec{a}\,[gradeup[\vec{a}]] = \vec{a}\,[\langle 1\ 2\ 0\ 3 \rangle] = \langle 2\ 0\ 1\ 3 \rangle[\langle 1\ 2\ 0\ 3 \rangle] = \langle 0\ 1\ 2\ 3 \rangle.$$

To clarify this example, we state the operations in words. The 0-th element of the index vector is 1, implying that the element in position 1 of the vector $\vec{a}$, i.e. 0, should be placed in the 0-th position of the result. The 1-st element of the index vector, 2, implies that the 2-nd element of $\vec{a}$, i.e. 1 should be placed in the 1-st position of the result and so on. We are now ready to define a general transpose for n-dimensional arrays.

Definition 5. *Given an array A with shape $\vec{s}$ such that the total number of components in $\vec{s}$ denotes the dimensionality d, of A, $A^{T\vec{i}}$ is defined whenever the shape of $A^{T\vec{i}}$ is $\vec{s}\left[\vec{t}\,\right]$, i.e. $\rho A^{T\vec{i}} = \vec{s}\left[\vec{t}\,\right]$. Then, for all $0 \le^* \vec{i} <^* \vec{s}\left[\vec{t}\,\right]$ (the symbols $\le^*$ and $<^*$ imply element by element comparisons):*

$$\vec{i}\psi A^{T\vec{i}} = \vec{i}\,\left[gradeup\left[\vec{t}\,\right]\right]\psi A$$

Example. Given

$$A = \begin{bmatrix} \begin{bmatrix} 0 & 1 & 2 \\ 3 & 4 & 5 \\ 6 & 7 & 8 \\ 9 & 10 & 11 \end{bmatrix} & \begin{bmatrix} 20 & 21 & 22 \\ 23 & 24 & 25 \\ 26 & 27 & 28 \\ 29 & 30 & 31 \end{bmatrix} \end{bmatrix}$$

We first look at $A^{T\langle 2\,1\,0\rangle}$ and note that this is equivalent to A^T. The shape of A is $\langle 2\ 4\ 3\rangle$ so the shape of $A^{T\langle 2\,1\,0\rangle}$ is $\langle 2\ 4\ 3\rangle[\langle 2\ 1\ 0\rangle] = \langle 3\ 4\ 2\rangle$. Then for all $0 \le^* \langle i\ j\ k\rangle <^* \langle 3\ 4\ 2\rangle$ (this is a shorthand notation for $0 \le i < 3; 0 \le j < 4;$ $0 \le k < 2$) we have:

$$\langle i\ j\ k\rangle\psi A^{T\langle 2\,1\,0\rangle} = (\langle i\ j\ k\rangle[gradeup[\langle 2\ 1\ 0\rangle]])\psi A$$
$$= (\langle i\ j\ k\rangle[\langle 2\ 1\ 0\rangle])\psi A$$
$$= \langle k\ j\ i\rangle\psi A \tag{8}$$

$$= \begin{bmatrix} \begin{bmatrix} 0 & 20 \\ 3 & 23 \\ 6 & 26 \\ 9 & 29 \end{bmatrix} & \begin{bmatrix} 1 & 21 \\ 4 & 24 \\ 7 & 27 \\ 10 & 30 \end{bmatrix} & \begin{bmatrix} 2 & 22 \\ 5 & 25 \\ 8 & 28 \\ 11 & 31 \end{bmatrix} \end{bmatrix} \tag{9}$$

Now let's look at another permutation of A noting there are 6 possible permutations, i.e. $\langle 0\ 1\ 2\rangle, \langle 0\ 2\ 1\rangle, \langle 1\ 2\ 0\rangle, \langle 1\ 0\ 2\rangle, \langle 2\ 0\ 1\rangle$, and $\langle 2\ 1\ 0\rangle$. This time let's look at $A^{T\langle 2\,0\,1\rangle}$. Now the shape of $A^{T\langle 2\,0\,1\rangle}$ is $\langle 2\ 4\ 3\rangle[\langle 2\ 0\ 1\rangle] = \langle 3\ 2\ 4\rangle$. Then for all $0 \le^* \langle i\ j\ k\rangle <^* \langle 3\ 2\ 4\rangle$

$$\langle i\ j\ k\rangle\ \psi\ A^{T\langle 2\,0\,1\rangle} = (\langle i\ j\ k\rangle[gradeup[\langle 2\ 0\ 1\rangle]])\psi A$$
$$= (\langle i\ j\ k\rangle[\langle 1\ 2\ 0\rangle])\psi A$$
$$= \langle j\ k\ i\rangle\psi A \tag{10}$$
$$= \begin{bmatrix} \begin{bmatrix} 0 & 3 & 6 & 9 \\ 20 & 23 & 26 & 29 \end{bmatrix} & \begin{bmatrix} 1 & 4 & 7 & 9 \\ 21 & 24 & 27 & 30 \end{bmatrix} & \begin{bmatrix} 2 & 5 & 8 & 11 \\ 22 & 25 & 28 & 31 \end{bmatrix} \end{bmatrix}$$

4   Changing Layouts Using Permutations

Now that we know how to permute an array over any of it's dimensions we can reorient the MoA outer product to have the same layout as the Kronecker product.

Recall the layouts of the Kronecker product in Eq. 5 and the MoA outer product in Eq. 6. Let's first permute the MoA outer product such that it has the same layout as the Kronecker product, and study the 4-d array defined by the MoA outer product in Eq. 4. Now observe the permuted array given in the following.

$$\begin{bmatrix} \begin{bmatrix} \begin{bmatrix} 1{\times}5 & 1{\times}6 & 1{\times}7 & 1{\times}8 \\ 2{\times}5 & 2{\times}6 & 2{\times}7 & 2{\times}8 \end{bmatrix} & \begin{bmatrix} 1{\times}9 & 1{\times}10 & 1{\times}11 & 1{\times}12 \\ 2{\times}9 & 2{\times}10 & 2{\times}11 & 2{\times}12 \end{bmatrix} & \begin{bmatrix} 1{\times}13 & 1{\times}14 & 1{\times}15 & 1{\times}16 \\ 2{\times}13 & 2{\times}14 & 2{\times}15 & 2{\times}16 \end{bmatrix} \end{bmatrix} \\ \begin{bmatrix} \begin{bmatrix} 3{\times}5 & 3{\times}6 & 3{\times}7 & 3{\times}8 \\ 4{\times}5 & 4{\times}6 & 4{\times}7 & 4{\times}8 \end{bmatrix} & \begin{bmatrix} 3{\times}9 & 3{\times}10 & 3{\times}11 & 3{\times}12 \\ 4{\times}9 & 4{\times}10 & 4{\times}11 & 4{\times}12 \end{bmatrix} & \begin{bmatrix} 3{\times}13 & 3{\times}14 & 3{\times}15 & 3{\times}16 \\ 4{\times}13 & 4{\times}14 & 4{\times}15 & 4{\times}16 \end{bmatrix} \end{bmatrix} \end{bmatrix} \tag{11}$$

Flattening the 4-d array of Eq. 11 gives us the layout we want. In other words, by flattening the array of Eq. 11 gives us the same one-dimensional array as obtained by flattening the Kronecker product (see Eq. 5).

Notice which dimensions changed between the initial outer product in Eq. 4 and the transposed outer product in Eq. 11. The shape went from $2 \times 2 \times 3 \times 4$ to $2 \times 3 \times 2 \times 4$. Reviewing Eqs. 5 and 6 we want 1 times 5, 6, 7, and 8 to be next to 2 times 5, 6, 7, and 8, etc. in the layout. Thus, we want to leave the 0th dimension alone, the 3rd dimension alone and we wanted to permute the 1st dimension with the 2nd. Consequently, we want $(A \; op_\times B)^{T\langle 0\,2\,1\,3\rangle}$, i.e. the $\langle 0\,2\,1\,3\rangle$ transpose of the outer product of A and B. Notice that this is the same permutation used in correlating the indices of A and B with the indices of the Kronecker product, i.e. $i, j, k, l \rightarrow i, k, j, l$. Recall the discussion in the paragraph following Eq. 3.

We now can discuss how to optimize these computations. Using MoA and ψ-Calculus, one can not only *compose* multiple indices in an array expression but, the algebraic reformulation of an expression can include processor/memory hierarchies. This is done by increasing the dimensions of the arguments. Through various restructurings, an expression can easily describe how to scale and port across complex processor/memory architectures.

The resultant matrix of the Kronecker product is traditionally evaluated and indexed as follows. The permutations on the input matrices in conjunction with an equivalent permutation on the corresponding shapes followed by a pairwise multiplication determines not only the resultant shape but how to store the results in its associated index of the resultant array. This cumbersome computation and encoding into new 2-d indices gets more and more complicated as the number of successive Kronecker products increases. Moreover, issues of parallelization complicate the problem since various components in the left argument are used over the columns of the result, assuming the partitioning was done by rows. Other partitions are possible: blocks, columns, etc. When the input matrices are large the problem is further complicated. This is not the case in MoA and ψ-Calculus.

5   Multiple Kronecker Products

Multiple Kronecker products are common in conjunction with quantum computing and high energy physics (e.g. *space-time* $\otimes$ *color* $\otimes$ *spin*). How can these be optimized to use basic abstract machine instructions (*start, stop, stride, count*) at all levels up and down the processor/memory hierarchy?

Presently, multiple Kronecker products require the materialization of each product. Notice what happens. After each pair of matrices is multiplied together, the result must be stored using the permutations of the indices of the argument arrays and encoded into row/column coordinates in a new matrix with size equal to the product of the pairs of permuted shapes. For example, if the input arrays were 2×2 and 3×3. The resultant shape would be a $(2 * 3) \times (2 * 3)$, i.e. 6×6. *At first thought, this looks fine since each pair can be parallelized.* Now, if we then did a Kronecker product with a 2×2 array, the result would be a 12×12 matrix. With each subsequent Kronecker product we'd need to store the product in the rows and columns associated with the permuted indices. Ideally, we want to *compose* multiple products in terms of their indexing such that we *index* each array once while eliminating temporary arrays. MoA and ψ-Calculus are ideally suited for this purpose and easily facilitate not only the composition of multiple Kronecker/outer products but their mapping to complex processor memory hierarchies.

To illustrate, let A be a 2×2 matrix and B be a 3×3 matrix. We are not concerned with the specific values of the matrix elements since we need only to consider manipulations of the indices. We assume the arithmetic is correctly defined. We'll perform $E = (A \otimes B) \otimes A$. The result within the parentheses would have shape 6×6. This was due to the two input array shapes, i.e. 2×2 and 3×3. Using, i, j to index A and k, l to index B bounded by their associated shapes, the composite index $\{i, k\}$ indexes the rows of $(A \otimes B)$ while the composite index $\{j, l\}$ indexes the columns of $(A \otimes B)$. For example, suppose we have arrays A and B given by:

$$A = \begin{bmatrix} 1 & 2 \\ 3 & 4 \end{bmatrix}, \tag{12}$$

and,

$$B = \begin{bmatrix} 0 & 1 & 2 \\ 3 & 4 & 5 \\ 6 & 7 & 8 \end{bmatrix}, \tag{13}$$

their Kronecker product is:

$$C = A \otimes B = \begin{bmatrix} 0 & 1 & 2 & 0 & 2 & 4 \\ 3 & 4 & 5 & 6 & 8 & 10 \\ 6 & 7 & 8 & 12 & 14 & 16 \\ 0 & 3 & 6 & 0 & 4 & 8 \\ 9 & 12 & 15 & 12 & 16 & 20 \\ 18 & 21 & 24 & 24 & 28 & 32 \end{bmatrix}. \tag{14}$$

Now let's compute $E = C \otimes A$. The result is:

$$E = \begin{bmatrix} 0 & 0 & 1 & 2 & 2 & 4 & 0 & 0 & 2 & 4 & 4 & 8 \\ 0 & 0 & 3 & 4 & 6 & 8 & 0 & 0 & 6 & 8 & 12 & 16 \\ 3 & 6 & 4 & 8 & 5 & 10 & 6 & 12 & 8 & 16 & 10 & 20 \\ 9 & 12 & 12 & 16 & 15 & 20 & 18 & 24 & 24 & 32 & 30 & 40 \\ 6 & 12 & 7 & 14 & 8 & 16 & 12 & 24 & 14 & 28 & 16 & 32 \\ 18 & 24 & 21 & 28 & 24 & 32 & 36 & 48 & 42 & 56 & 48 & 64 \\ 0 & 0 & 3 & 6 & 6 & 12 & 0 & 0 & 4 & 8 & 8 & 16 \\ 0 & 0 & 9 & 12 & 18 & 24 & 0 & 0 & 12 & 16 & 24 & 32 \\ 9 & 18 & 12 & 24 & 15 & 30 & 12 & 24 & 16 & 32 & 20 & 40 \\ 27 & 36 & 36 & 48 & 45 & 60 & 36 & 48 & 48 & 64 & 60 & 80 \\ 18 & 36 & 21 & 42 & 24 & 48 & 24 & 48 & 28 & 56 & 32 & 64 \\ 54 & 72 & 63 & 84 & 72 & 96 & 72 & 86 & 84 & 112 & 96 & 128 \end{bmatrix} \tag{15}$$

Recall that the result matrix is filled in by 6, 2×2 blocks, over the rows and columns using the encoding discussed above. Notice how complicated the indirect addressing becomes using this approach to implementation of the Kronecker product. Notice also that if we wanted to distribute the computation of a block of rows to 4 processors, we'd need multiple components of the left argument.

Let us now look at doing the same operations, i.e. multiple outer products, using the MoA, ψ-Calculus approach. We find:

$$C = A\,op_\times B =$$

$$\left[\left[\begin{bmatrix} 0\ 1\ 2 \\ 3\ 4\ 5 \\ 6\ 7\ 8 \end{bmatrix} \begin{bmatrix} 0\ \ 2\ \ 4 \\ 6\ \ 8\ 10 \\ 12\ 14\ 16 \end{bmatrix}\right]\left[\begin{bmatrix} 0\ \ 3\ \ 6 \\ 9\ 12\ 15 \\ 18\ 21\ 24 \end{bmatrix} \begin{bmatrix} 0\ \ 4\ \ 8 \\ 12\ 16\ 20 \\ 24\ 28\ 32 \end{bmatrix}\right]\right], \tag{16}$$

is a 4-d array with shape $\langle 2\,2\,3\,3\rangle$. It is easy to see that indexing this array with *partial indices* yields 4, 3×3 sub-arrays. That is, the indices, $\langle 0\,0\rangle$, $\langle 0\,1\rangle$, $\langle 1\,0\rangle$ and $\langle 1\,1\rangle$ are used to index C and each sub-array would be sent to available processors $0, 1, 2$, and 3, to create a *start, stop, stride*, mapping suitable for all architectures to date. An example of selecting a subarray with a partial index is given by:

$$\langle 1\,0\rangle\psi C = \begin{bmatrix} 0\ \ 3\ \ 6 \\ 9\ 12\ 15 \\ 18\ 21\ 24 \end{bmatrix}. \tag{17}$$

Mapping to Processors and the ONF. Next consider the operation $C\,op_\times A$. This would yield a 6-*d* array with shape $\langle 2\,2\,3\,3\,2\,2\rangle$. We can easily pull apart the arguments in the operations. We can also *reshape* this array into one of shape $\langle 4\,3\,3\,2\,2\rangle$. We then would use the leftmost index to denote the processors over which the array has been partitioned. We know the blocks each have 36 components, i.e. the product of the shape exclusive of the first component: $3 \times 3 \times 2 \times 2$.

The following expressions illustrate how easy it is to compose, map, and scale to a multi-processor architecture. As before, we get the shape.

$$\rho((A\ op\ B)\ op\ A) = (\rho(A\ op\ B) +\!\!+ (\rho A))$$
$$= (\rho A) +\!\!+ (\rho B) +\!\!+ (\rho A) \tag{18}$$
$$= \langle 2\,2\,3\,3\,2\,2\rangle$$

and compose the indices.

Given $0 \leq^* \langle i\ j\rangle <^* \langle 2\ 2\rangle;\ 0 \leq^* \langle k\ l\rangle <^* \langle 3\ 3\rangle;\ 0 \leq^* \langle m\ n\rangle <^* \langle 2\ 2\rangle$ and for all $0 \leq^* \langle i\ j\ k\ l\ m\ n\rangle <^* \langle 2\,2\,3\,3\,2\,2\rangle$;

$$\langle i\ j\ k\ l\ m\ n\rangle\psi\,((A op_\times B)\ op_\times\ A)$$
$$= ((\langle i\ j\ k\ l\rangle\psi\,(A\ op_\times\ B)) \times (\langle m\ n\rangle\psi A) \tag{19}$$
$$= ((\langle i\ j\rangle\psi A) \times (\langle k\ l\rangle\psi B) \times (\langle m\ n\rangle\psi A)$$

From here we can easily map subarrays to the four processors using *starts, stops,* and *strides*.

Let's take the above, referred to as the *Denotational Normal Form* (DNF) expressed in terms of *Cartesian coordinates* and transform it into its equivalent *Operational Normal Form* (ONF) expressed in terms of *start, stop, stride* and *count*. The DNF is independent of layout. The ONF requires one. Let's assume row-major (although column major or other orderings are equally possible). We'll see how natural that is for the outer product at all levels of implementation.

Let's break up the above multiple outer product over 4 processors. We'll need to restructure the array's shape $\langle 2\ 2\ 3\ 3\ 2\ 2 \rangle$, to $\langle 4\ 3\ 3\ 2\ 2 \rangle$. This allows us to index the 0-th (i.e. the *leftmost*) dimension of this abstraction over the processors. The following equation shows the result of this partitioning.

Under this partitioning, equation 19 for $0 \leq p < 4$, becomes:

$$\langle i\ j\ k\ l\ m\ n \rangle \psi\ ((A\,op_\times B)\ op_\times A)$$
$$= \langle p\ k\ l\ m\ n \rangle \psi\ ((\vec{a}\ op_\times B)\ op_\times A \qquad (20)$$
$$= ((\langle p \rangle \psi\, \vec{a}) \times (\langle k\ l \rangle \psi B)) \times (\langle m\ n \rangle \psi A),$$

where $\langle p \rangle \psi \vec{a}$ denotes the value located at the address $@A[p]$ (i.e. the address of A offset by p).

The entire right argument of Eq. 20 is used/accessed in all of the processors. Thus, we think of the entire result of both products residing in an array with $\pi\langle 4\ 3\ 3\ 2\ 2 \rangle = 144$ components (the π operator gives the product of the elements of the vector) laid out contiguously in memory using a row-major ordering.

Thus, the equation above for $0 \leq p < 4$, becomes:

$$\langle i\ j\ k\ l\ m\ n \rangle \psi\ ((A\,op_\times B)\ op_\times A)$$
$$= \langle p\ k\ l\ m\ n \rangle \psi\ ((\vec{a}\ op_\times B)\ op_\times A \qquad (21)$$
$$= ((\langle p \rangle \psi\, \vec{a}) \times (\langle k\ l \rangle \psi B)) \times (\langle m\ n \rangle \psi A),$$

where $\langle p \rangle \psi \vec{a}$ denotes $@A[p]$, i.e. the address of A offset by p.

Pseudo Code from the ONF. The following *pseudo-code* expression describes what each processor, p, will do.
$$\forall\, p, q, r\ \ s.t.\ 0 \leq p < 4\ ;\ 0 \leq q < 9;\ 0 \leq r < 4$$

```
(avec[p] * bvec[q]) * avec[r]
```

In the above, **avec** and **bvec** are used to describe generic implementations. We are able to collapse the 2-*d* indexing for A and B since their access is contiguous. The approach we have just outlined has been applied to a number of other problems [17, 18, 21].

6 n Kronecker Products: A New Theorem

We saw in previous sections how the Kronecker Product *remained* a *matrix* and how that fact may pose problems in parallelizing the executions. In applications such as high energy physics, pairs of Kronecker Products are computed independently on huge complex processor memory hierarchies. Imagine, if we could *compose* the Kronecker Products and determine *prior to compilation*, how to eliminate each temporary formed by each pair of products. Moreover, imagine if we could *easily* determine how to map these operations due to the contiguity of the data structure. We also saw in previous sections that we could *permute* the indices of the Kronecker Product to exploit such locality. So far we've shown how to compose *three* outer products and permute the result such that it is a matrix in a Kronecker Product layout. ***We now show that this result generalizes to any number of outer products resulting in an n-dimensional array.***

Creating the Transpose Vector for an n-dimensional Transpose. We previously saw that when we did *two outer products* we had to permute the resultant 4-dimensional array using a $\langle 0\ 2\ 1\ 3 \rangle$ transpose vector. If we were to compose n Kronecker Products we would create an $n * 2$ dimensional array result. Thus, in order to permute the indices such that we have a Kronecker Product layout we would create a vector where the first half contained the even numbers from 0 to $(2 \times n) - 2$ and the second half of the vector contained the odd numbers from 1 to $(2 \times n) - 1$. This vector would permute each index of the resultant array by that permutation. Then, we'd reshape the array back into a two dimensional array. This conjecture is easily proved by induction based on the properties of the generalized transpose operator and the group properties of permutations of indices.

Example. Let's continue with our original example with a resultant array of $\langle 2\ 2\ 3\ 3\ 2\ 2 \rangle$. This array denotes 3 outer products and is a 6-dimensional array. Thus, the transpose vector would be $\langle 0\ 2\ 4\ 1\ 3\ 5 \rangle$. We then get the product of the shape of the even indices and odd indices. In this case it is $\langle 12\ 12 \rangle$. Consequently, the resultant array after optimization is:

$$\langle 12\ 12 \rangle \text{reshape}(\langle 0\ 2\ 4\ 1\ 3\ 5 \rangle \text{transpose} A')$$

where A' is the 6-dimensional result of the composed and optimized MoA outer products.

Implement These Ideas in MATLAB. As a first goal, we wish to be able to call compiled programs from MATLAB, but eventually we hope to identify an algebraic subset of MATLAB that is equivalent to the MoA algebra. Then, using a compiler flag, users could opt to develop their MATLAB programs using the subset to create flexible, high-performance programs that could also be called from standard MATLAB. By incorporating these concepts into production level programming languages we believe the user community will gradually adopt the subset and thus create numerous callable high-performance routines. If we can identify an algebraic subset of all array languages that are equivalent to the MoA algebra, they could all be mapped to each other: Fortran95, Python, etc. This promises to allow interpretive languages to perform at the speed of compiled programs.

Progress in Implementation: An "Abstract Machine" for Tensor Operations. In order to streamline the computation of multiple Kronecker Products we wish to take advantage of data locality. As such we have derived an algorithm that enables us to compute an array of indicies. This array of indices describes where a particular element of the multiple Kronecker Product should be stored. Having such an array of indices simplifies the computation by allowing us to cycle through the computation of the of the "right hand sides" in a contiguous fashion and assign whole blocks of the result to the result vector.

Our algoritm for the index vector shows that that index vector can be seen to be of the form of an "outer product plus." In particular we have found a monolithic MoA expression for the index vector. An in-depth explanation of the derivation will not be attempted. Rather we will explain the result in some brief detail.

For a concrete example we consider a multiple Kronecker Product of four 4×4 matrices. Given a multi-dimensional index for the result of the form $\langle i j k l m n o p \rangle$, one can understand the resultant offset as a polynomial expression involving the elements of the index vector times various products of elements of the shape vector. Without specifying in detail the values of these shape vector products, let us denote them by a vector of values $\langle c_0 \, c_1 \, c_2 \, c_3 \, c_4 \, c_5 \, c_6 \, c_7 \rangle$.

Next we form a collection of vectors $\vec{d}_m$ with $0 \leq m < 8$ such that $\vec{d}_m$ is a product of the scalar c_m with the vector of integers $\langle 0\,1\,2\,3 \rangle$, so that $\vec{d}_m \equiv c_m \times \langle 0\,1\,2\,3 \rangle = \langle 0 \, c_m \, 2c_m \, 3c_m \rangle$.

We next form the arrays "outer product plus" arrays:

$$B_0 = \vec{d}_0 \; op_+ \; \vec{d}_1$$

$$B_1 = \vec{d}_2 \; op_+ \; \vec{d}_3$$

$$B_2 = \vec{d}_4 \; op_+ \; \vec{d}_5$$

$$B_3 = \vec{d}_6 \; op_+ \; \vec{d}_7$$

The final result for the monolithic index vector is given by:

$$res = B_0 \; op_+ \; B_1 \; op_+ \; B_2 \; op_+ \; B_3.$$

The first few values of this array, when flattened into a vector are as follows:

$$\langle 0 \; 1 \; 2 \; 3 \; 256 \; 257 \; 258 \; 259 \; 512 \; 513 \; 514 \; 515 \; 768 \; 769 \; 770 \; 771 \; 4 \; 5 \; 6 \; 7 \cdots \rangle.$$

Note that this index vector is contiguous in blocks of four. If we were working with outer products of larger arrays, the contiguous blocks would be longer in length. Our derivation is completely deterministic in the sense that for any number of outer products, we can **predict** not only the length of the contiguous blocks but also the values in each block.

Understanding the algorithm in depth allows us to take full advantage of data locality so as to be able to correctly prefetch blocks of data on the left and right side of the computation so as to effectively overlap computation and IO.

Implementation Using Low-Level SIMD Instructions. We have begun computationally implementing multiple Kronecker Products in C using low level SIMD instructions provided for Intel Processors. Subroutines written in this way were made callable by MATLAB (and Octave) using the commonly used "void mex-Function" interface.

The basic data type used in these implementations is the so-called vFloat. A vFloat variable is a 128 bit register variable representing four floats. Alternatively a vFloat variable can be used to represent two double precision numbers. Also available are arrays of vFloat variables. Thus to represent N floats, we use a length $N/4$ vFloat array. Thus we see that, given the SIMD hardware instructions, we are necessarily constrained to work with arrays that are of length N which is a multiple of 4.

Initial results are promising. We have compared the result of our multiple Kronecker Product routine, implemented using SIMD instructions to what one would obtain using MATLAB or Octave's compiled "kron" subroutine. We ran all tests using Octave to compute three Kronecker Products of four, 4×4 matrices and found a factor of 2 speedup as compared to simply calling "kron" three times.

It is important to take a moment to reflect on this result. Note, in our SIMD implementation, we have combined all three Kronecker Products into one operation. In effect do a multiple loop over all possible values of the product. In contrast, calling "kron" multiple times does effectively fewer computational steps but is **more time consuming because of the cost of the intermediate temporaries**. To be more precise, consider a Kronecker Product of two $m \times m$ matrices which each contain $n = m^2$ values. To compute this in the conventional way requires n^2 operations. Thus computing three such Kronecker Products requires $3n^2$ operations. In contrast by combining all three into a single operation leads to nested loops resulting in effectively n^4 operations.

Given this analysis, how could the MoA multiple Kronecker Product be twice as fast as the conventional approach? The answer is that we, by using the SIMD instructions along with the monolithic computation of the index vector, are more effectively utilizing data locality to overlap computation and IO. We are also getting a computational cost savings by eliminating intermediate temporary arrays.

These initial results are encouraging and have motivatated continuing development efforts along these lines. Further work needs to be done to develop a completely general function that arbitrary input arrays and computes an arbitrary number of Kronecker Products of them. Initial steps along these lines will focus on the practical problem of input arrays comprising square matrices. The general case can, of course, be handled by MoA techniques although such a general case is considerably more complicated.

7   Conclusion

Optimization using the techniques of MoA and the ψ-Calculus is straightforward and can be done at the time of parsing and tokenization. Once the syntax is removed and tokens are used to create the execution tree, which is used by the compiler, the tree is walked multiple times. First the shapes of each expression are determined, then from the shapes the tree is walked again to translate all formal operations into expressions involving direct indexing. The resultant tree is the DNF. From that, the mapping function γ (that takes a shape and an index and returns an offset) is used to map Cartesian coordinates to starts, stops, and strides as illustrated in this paper. This can be directly compiled to hardware.

We have illustrated this approach in the context of efficient computation of multiple Kronecker Products and have presented a discussion of its implementation using low level SIMD instructions. We have implemented a SIMD based multiple Kronecker Product routine that is compiled and callable by MATLAB or Octave. Initial results of numerical tests are promising, with a speedup of a factor of two over the use of Octave's built in "kron" program to compute three Kronecker Products of four matrices.

References

1. Abrams, P.S.: What's wrong with APL? In: APL 1975: Proceedings of Seventh International Conference on APL, pp. 1–8. ACM, New York (1975)
2. Abrams, P.S.: An APL machine. PhD thesis, Stanford University, Stanford, CA, USA (1970)
3. Acar, E., Dunlavy, D.M., Kolda, T.G., Morup, M.: Scalable tensor factorizations with missing data. In: SDM 2010: Proceedings of the 2010 SIAM International Conference on Data Mining (April 2010)
4. Acar, E., Kolda, T.G., Dunlavy, D.M.: An optimizations approach for fitting canonical tensor decompositions. Technical Report SAND2009-0857, Sandia National Laboratories, Albuquerque, NM and Livermore, CA (February 2009)
5. Bader, B.W., Kolda, T.G.: Matlab tensor toolbox version 2.4 (2001), http://csmr.ca.sandia.gov/~tgkolda/TensorToolbox/
6. Bader, B.W., Kolda, T.G.: Algorithm 862: Matlab tensor classes for fast algorithm prototyping. ACM Transactions on Mathematical Software 32(4) (December 2006)
7. Bader, B.W., Kolda, T.G.: Efficient matlab computations with sparse and factored matrices. SIAM Journal on Scientific Computing 30(1), 205–231 (2007)
8. Berkling, K.: Arrays and the lambda calculus. Technical report, CASE Center and School of CIS, Syracuse University (1990)
9. Eatherton, W., Kelly, J., Schiefelbein, T., Pottinger, H., Mullin, L.R., Ziegler, R.: An fpga based reconfigurable coprocessor board utilizing a mathematics of arrays. Technical report, University of Missouri–Rolla, Computer Science Department (1995)
10. Gerhart, S.: Verification of APL Programs. PhD thesis, CMU (1972)
11. Helal, M.A.: Dimension and shape invariant programming: The implementation and the application. Master's thesis, The American University in Cairo, Department of Computer Science (2001)
12. Kennedy, A., et al.: (October 2009), http://kac.maths.ed.ac.uk/nsf-nais/home.php
13. Kepner, J.: Parallel matlab for multicore and multinode camputers. SIAM, Philadelphia (2009)

14. Kolda, T.G., Bader, B.W.: Tensor decompositions and applications. SIAM Review 51(3), 455–500 (2009)
15. McMahon, T.: Mathematical formulation of general partitioning of multi-dimensional arrays to multi-dimensional architectures using the Psi calculus. Undergraduate Honors Thesis (1995)
16. Mermin, N.D.: Quantum Computer Science. Cambridge University Press, Cambridge (2007)
17. Mullin, L.M.R.: A Mathematics of Arrays. PhD thesis, Syracuse University (December 1988)
18. Mullin, L.R.: A uniform way of reasoning about array–based computation in radar: Algebraically connecting the hardware/software boundary. Digital Signal Processing 15, 466–520 (2005)
19. Mullin, L., Kluge, W., Scholtz, S.: On programming scientific applications in SAC – a functional language extended by a subsystem for high level array operations. In: Kluge, W.E. (ed.) IFL 1996. LNCS, vol. 1268, pp. 85–104. Springer, Heidelberg (1997)
20. Mullin, L., Nemer, N., Thibault, S.: The Psi compiler v4.0 for HPF to Fortran 90: User's Guide. Department of Computer Science, University of Missouri–Rolla (1994)
21. Mullin, L.R., Raynolds, J.E.: Conformal computing: Algebraically connecting the hardware/software boundary using a uniform approach to high-performance computation for software and hardware applications. CoRR, abs/0803.2386 (2008)
22. Mullin, L., Rutledge, E., Bond, R.: Monolithic compiler experiments using C++ expression templates. In: Proceedings of the High Performance Embedded Computing Workshop (HPEC 2002). MIT Lincoln Lab, Lexington (2002)
23. Mullin, L., Rutledge, E., Bond, R.: Monolithic compiler experiments using C++ Expression Templates. In: Proceedings of the High Performance Embedded Computing Workshop HPEC 2002. MIT Lincoln Laboratory, Lexington (2002)
24. Mullin, L., Thibault, S.: Reduction semantics for array expressions: The Psi compiler. Technical Report CSC 94-05, Department of Computer Science, University of Missouri-Rolla (1994)
25. NSF-NAIS Workshop Intelligent Software: The Interface between Algorithms and Machines, Ediburgh, Scotland (October 2009),
    http://adrg.maths.ed.ac.uk/nsf-nais
26. Pottinger, H., Eatherton, W., Kelly, J., Schiefelbein, T., Mullin, L.R., Ziegler, R.: Hardware assists for high performance computing using a mathematics of arrays. In: FPGA 1995: Proceedings of the 1995 ACM Third International Symposium on Field-Programmable Gate Arrays, pp. 39–45. ACM, New York (1995)
27. Raynolds, J.E., Mullin, L.R.: Applications of conformal computing techniques to problems in computational physics: the fast fourier transform. Computer Physics Communications 170(1), 1–10 (2005)
28. Rosenkrantz, D.J., Mullin, L.R., Hunt III, H.B.: On minimizing materializations of array-valued temporaries. ACM Trans. Program. Lang. Syst. 28(6), 1145–1177 (2006)
29. Tu, H.-C.: FAC: A Functional Array Calculator and it's Applicaton to APL and Functional Programming. PhD thesis, Yale University (1985)
30. Tu, H.-C., Perlis, A.J.: FAC: A functional APL language. IEEE Software 3(1), 36–45 (1986)
31. Van Loan, C.: (February 2009),
    http://www.cs.cornell.edu/cv/tenwork/home.htm

32. Van Loan, C.: (May 2009),
    http://www.cs.cornell.edu/cv/tenwork/finalreport.pdf
33. Van Loan, C.: The Kronecker product: A product of the times. In: SIAM Conference on Applied Linear Algebra, Monterey, California (October 2009)

Reliable and Robust Automated Synthesis of QFT Controller for Nonlinear Magnetic Levitation System Using Interval Constraint Satisfaction Techniques

P.S.V. Nataraj* and Mukesh D. Patil

Department of Systems and Control Engineering,
Indian Institute of Technology, Bombay, Mumbai, 400 076. India
{nataraj,mdpatil}@sc.iitb.ac.in
http://www.sc.iitb.ac.in

Abstract. Robust controller synthesis is of great practical interest and its automation is a key concern in control system design. Automatic controller synthesis is still a open problem. In this paper a new, efficient method has been proposed for automated synthesis of a fixed structure quantitative feedback theory (QFT) controller by solving QFT quadratic inequalities of robust stability and performance specifications. The controller synthesis problem is posed as interval constraint satisfying problem (ICSP) and solved with interval constraint solver (realpaver)[1] . The method is guaranteed to find all feasible controllers of given structure in the search domain. The controller designed using proposed method is experimentally tested on ECP's Magnetic Levitation system [2] which has open loop stable and unstable configurations.

Keywords: Robust Control, QFT, Automatic Loop Shaping, Constraint Propagation, Interval Analysis, ICSP.

1 Introduction

A key step in the quantitative feedback theory (QFT) approach to robust control system design (see [3]) is the one of synthesizing the controller. In this step, a controller is synthesized to satisfy the magnitude-phase QFT *bounds* on the nominal loop transmission function at each design frequency. Traditionally, this synthesis was done *manually* by the designer, relying on design experience and skill. Recently, several researchers have attempted to *automate* this step, see, for instance, [4–6, 9–11].

The concept of controller design automation in QFT was introduced by [6] who proposed an iterative procedure based on Bode's famous gain-phase integral to derive the shape of a nominal loop transfer function. The method, however, needs a rational function approximation to obtain an analytical expression for

---

* Corresponding author.

M. Ceberio and V. Kreinovich (eds.), *Constraint Programming and Decision Making*,
Studies in Computational Intelligence 539,
DOI: 10.1007/978-3-319-04280-0_15, © Springer International Publishing Switzerland 2014

the loop transfer function, and straight line approximations for the nonlinear QFT bounds.

Thomspon and Nwokah proposed a method based on nonlinear programming techniques wherein the templates of the uncertain plant are approximated by overbounding rectangles. Such a template approximation leads to overbounding in the constraints derived for the optimization.

Bryant and Halikias addressed the problem of automatic loop shaping using linear programming techniques wherein the QFT bounds are approximated by a series of linear approximations. However, their method also leads to conservatism in describing nonlinear QFT bounds by linear inequalities.

Chait *et al* proposed a method based on convexification of the non-convex closed loop bounds [7]. The QFT design problem in this method is posed in terms of the closed loop complementary sensitivity function. In this method the closed loop non-convex bounds are transformed into linear inequalities without any conservatism, and then a linear program is solved. However, as pointed by these authors, the shortcoming of the method is that it involves fixing the poles of the closed loop transfer function *a priori*.

Synthesis of controller is treated as an optimization problem in Chen *et al.* [8–10] reformulated the problem as one of parameter optimization of a fixed structure controller.

To the best of our knowledge, no method for QFT fixed order controller synthesis for unstable configuration has been proposed using quadratic inequalities of robust stability and performance specifications. In this paper, an ICSP based method is proposed for automated synthesis of QFT controller for unstable configuration of Magnetic Levitation system.

2   Some Preliminaries

2.1   Quantitative Feedback Theory

Consider a two degree freedom feedback system configuration (see Fig 1), where $G(s)$ and $F(s)$ are the controller and prefilter respectively. The uncertain linear time-invariant plant $P(s)$ is given by $P(s) \in \{P(s, \lambda) : \lambda \in \boldsymbol{\lambda}\}$, where $\lambda \in \Re^l$ is a vector of plant parameters whose values vary over a parameter box $\boldsymbol{\lambda}$

$$\boldsymbol{\lambda} = \{\lambda \in \Re^l : \lambda_i \in [\underline{\lambda_i}, \overline{\lambda_i}], \ \underline{\lambda_i} \leq \overline{\lambda_i}, \ i = 1, ..., l\}$$

This gives rise to a parametric plant family or set

$$\mathcal{P} = \{P(s, \lambda) : \lambda \in \boldsymbol{\lambda}\}$$

The open loop transmission function is defined as

$$L(s, \lambda) = G(s)P(s, \lambda) \tag{1}$$

and the nominal open loop transmission function is

$$L_0(s) = G(s)P(s, \lambda_0) = l_0 \exp^{j\psi_0} \tag{2}$$

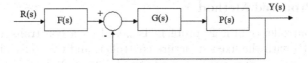

Fig. 1. The two degree-of-freedom structure in QFT

2.2 Quadratic Inequalities for Parametric Uncertainty

The quadratic inequalities corresponding to the stability and performance specifications for parametric uncertainties are [12, 13]:

- Robust stability specification,

$$g^2 p^2 + 2gp \cos(\phi + \theta) + 1 \geq 0 \tag{3}$$

- Robust stability margin specification,

$$g^2 p^2 \left(1 - \frac{1}{w_s^2}\right) + 2gp \cos(\phi + \theta) + 1 \geq 0 \tag{4}$$

- Robust sensitivity or (output disturbance rejection) specification,

$$g^2 p^2 + 2gp \cos(\phi + \theta) + \left(1 - \frac{1}{w_{do}^2(\omega)}\right) \geq 0 \tag{5}$$

- Tracking specification,

$$g^2 p_k^2 p_i^2 \left(1 - \frac{1}{\delta^2(\omega)}\right) + 2gp_k p_i \left[p_k \cos(\phi + \theta_i) - \frac{p_i}{\delta^2(\omega)} \cos(\phi + \theta_k)\right] + \left(p_k^2 - \frac{p_i^2}{\delta^2(\omega)}\right) \geq 0 \tag{6}$$

where

$$\delta(\omega) = \frac{|T_U(j\omega)|}{|T_L(j\omega)|}$$

3 Controller Synthesis Methodology

3.1 The Loop Shaping Problem

The QFT loop shaping problem for single-input single-output (SISO) or multi-input single-output (MISO) systems or the MIMO systems can be described as follows: Find a stabilizing linear time-invariant (LTI) controller $G(s)$ such that the feedback system whose nominal open-loop transmission function, $L_0(s, \lambda) = G(s)P_0(s, \lambda_0)$, satisfies

$$L(s, \lambda) = G(s)P(s, \lambda) \in B(w), \quad \forall \ w \geq 0$$

where, $B(w)$, for any frequency w, denotes a set in the complex plane.

3.2   The Proposed Method

The QFT controller synthesis problem is posed as a constrained satisfaction problem (CSP) with the fixed structure controller, and the constraint set as the set of possibly non-convex, nonlinear magnitude-phase QFT bounds for stability and performance specifications, at the various design frequencies. In this work, the constraints are solved using constraint solver Realpaver. Several consistency techniques like box, hull, 3B are implemented in Realpaver [14].

The controller synthesis procedures that lack an analytical or closed form solution are usually iterative in nature, involving trial-and-error techniques and/or thumb rules. The success of iterative controller synthesis process depends considerably upon the expertise of the designer. With the exponential growth and easy availability of computational power, such designs can now be automated.

The design automation can be posed as a constrained or unconstrained solving problem (CSP). The steps involved in this process would be as follows:

- Selection of the robust control methodology.
- Formulation of the synthesis problem:
  - Conversion of the control synthesis problem into ICSP
  - Choice of the controller structure.
  - Specification of the controller parameter search space.
- Choice of the ICSP solver for the above problem.

Algorithm for ICSP

Input: Constraints C, Initial Search Box, **B** and accuracy ϵ
Output: Solution Box with all feasible controllers or "NO Solution Exists "

1. Initialize the box list L with initial box B
2. Take a box from list L and prune it. If no box in list, Exit.
3. If box can not be pruned further and width of box is less or equal to ϵ, store the box as solution. Go to step 2
4. Bisect the box in maximum width direction and put the sub-boxes in list.
5. Go to step 2.

The key features of the proposed method are:

- It enables the designer to specify in advance the structure of the controller to be synthesized.
- It can deal directly with the numerical values of the possibly non-convex, nonlinear QFT bounds at each design frequency. The QFT bounds can be generated with, say, the QFT toolbox.
- It automatically takes care of the internal stability of the system.
- If for the specified structure and the given search box of controller parameter values,
  - no feasible controller exists, then the method is guaranteed to computationally verify this fact.
  - a feasible controller does exist, then the method is guaranteed to find all controllers lying within the search box.

References

1. Granvilliers, L., Benhamou, F.: RealPaver: An Interval Solver using Constraint Satisfaction Techniques. ACM Transaction on Mathematical Software 32, 138–156 (2006)
2. Manual for Model 730, Magnetic Levitation System. Educational Control Products, California, USA (1999)
3. Horowitz, I.M.: Quantitative feedback design theory (QFT). QFT Publications, Boulder (1993)
4. Bryant, G.F., Halikias, G.D.: Optimal loop-shaping for systems with large parameter uncertainty via linear programming. Int. J. Control 62(3), 557–568 (1995)
5. Chait, Y., Chen, Q., Hollot, C.V.: Automatic loop-shaping of QFT controllers via linear programming. Trans. of the ASME Journal of Dynamic Systems, Measurement and Control 121, 351–357 (1999)
6. Gera, A., Horowitz, I.M.: Optimization of the loop transfer function. Int. J. Control 31, 389–398 (1980)
7. Chait, Y., Yaniv, O.: MISO computer aided control design using QFT. International Journal of Robust and Nonlinear Control (1998)
8. Chen, W., Ballance, D.J., Li, Y.: Automatic loop-shaping in qft using genetic algorithms. In: Proceedings of 3rd Asia-Pacific Conference on Control and Measurement, pp. 63–67 (1998)
9. Garcia-Sanz, M., Guillen, J.C.: Automatic loop-shaping of QFT robust controllers via genetic algorithms. In: Proceedings of the 3rd IFAC Symposium on Robust Control Design, Kidlington, U.K. (2000)
10. Paluri, N.S.V., Tharewal, S.: An interval analysis algorithm for automated controller synthesis in QFT designs. Trans. of the ASME Journal of Dynamic Systems, Measurement and Control 129, 311–321 (2007)
11. Thomspon, D.F., Nwokah, O.D.I.: Analytical loop shaping methods in quantitative feedback theory. Trans. of the ASME Journal of Dynamic Systems, Measurement and Control 116, 169–177 (1994)
12. Chait, Y., Tsypkin, Y.: SISO QFT design with Non-parametric Uncertainties. Presentation at the 1993 American Control Conference. University of Massachusetts, USA, pp. 1694–1695 (1993)
13. Chait, Y., Yaniv, O.: Multi-input/single-output computer-aided control design using quantitative feedback theory. International Journal of Robust and Nonlinear Control 3(1), 47–54 (1993)
14. Benhamou, F., Goualard, F., Granvilliers, L.: Revising hull and box consistency. In: Proc. of 16th International Conference on Logic Programming, pp. 230–244 (1999)

Towards an Efficient Bisection of Ellipsoids

Paden Portillo, Martine Ceberio, and Vladik Kreinovich

Department of Computer Science,
University of Texas at El Paso, El Paso, TX 79968, USA
pportillo2@miners.utep.edu, {vladik,mceberio}@utep.edu

Abstract. Constraints are often represented as ellipsoids. One of the main advantages of such constrains is that, in contrast to boxes, over which optimization of even quadratic functions is NP-hard, optimization of a quadratic function over an ellipsoid is feasible. Sometimes, the area described by constrains is too large, so it is reasonable to bisect this area (one or several times) and solve the optimization problem for all the sub-areas. Bisecting a box, we still get a box, but bisecting an ellipsoid, we do not get an ellipsoid. Usually, this problem is solved by enclosing the half-ellipsoid in a larger ellipsoid, but this slows down the domain reduction process. Instead, we propose to optimize the objective functions over the resulting half-, quarter, etc., ellipsoids.

Keywords: constraints, ellipsoids, bisection, computational complexity.

Constraints on a Single Variable. In many practical problems, we have prior *constraint* on the values of different quantities. For each individual quantity x, we usually know a lower bound $\underline{x}$ and an upper bound. Thus, we know that the actual value of this quantity must lie within the interval $[\underline{x}, \overline{x}]$.

Sometimes, we know several lower bounds; in this case, we take the largest of them as $\underline{x}$. Similarly, if we know several upper bounds, we can take the smallest of these upper bounds as $\overline{x}$.

Correspondingly, when are looking for a value that satisfies a certain condition (e.g., when we are solving an equation), or if we are looking for the best option (i.e., solving an appropriate optimization problem), we should take this constraint into account. For example, when finding the optimal value of x, we should optimize the corresponding objective function $f(x)$ under the given constraint on x – i.e., under the constraint that $\underline{x} \le x \le \overline{x}$.

Constraints on Several Variables: Boxes Naturally Appear. Usually, we have several different variables $x_1, \ldots, x_n$. For each of these variables x_i, we usually know a lower bound $\underline{x}_i$ and an upper bound $\overline{x}_i$. Thus, we know that the actual value of the tuple $x = (x_1, \ldots, x_n)$ belongs to the *box* $[\underline{x}_1, \overline{x}_1] \times \ldots \times [\underline{x}_n, \overline{x}_n]$. Such box constrains and box uncertainty are typical for *interval computations*; see, e.g., [8, 9, 12].

M. Ceberio and V. Kreinovich (eds.), *Constraint Programming and Decision Making*,   137
Studies in Computational Intelligence 539,
DOI: 10.1007/978-3-319-04280-0_16, © Springer International Publishing Switzerland 2014

Problem with Box Constraints: Computational Complexity. The main problem with box constraints is that already for quadratic objective functions

$$f(x_1, \ldots, x_n) = a_0 + \sum_{i=1}^{n} a_i \cdot x_i + \sum_{i=1}^{n} \sum_{j=1}^{n} a_{ij} \cdot x_i \cdot x_j,$$

optimizing them over a box is, in general, an NP-hard problem; see, e.g., [11, 15]. This problem is computational complex not for some exotic quadratic function: as shown in [5], it is actually even NP-hard for the sample variance

$$V(x_1, \ldots, x_n) = \frac{1}{n} \cdot \sum_{i=1}^{n} x_i^2 - \left(\frac{1}{n} \cdot \sum_{i=1}^{n} x_i \right)^2.$$

Informal Explanation of Computational Complexity. The above computational complexity can be intuitively explained.

Indeed, a function of one variable $f(x)$ attains its optimum (maximum or minimum) on an interval $[\underline{x}, \overline{x}]$ either at one of its endpoints, or at an internal point $x \in (\underline{x}, \overline{x}]$. If this optimum is attained at an internal point, then at this point, a derivative $\dfrac{df}{dx}$ should be equal to 0. Thus, to find the largest and the smallest value of a function $f(x)$ on the interval $[\underline{x}, \overline{x}]$, it is sufficient to consider its value at the endpoints $\underline{x}$ and $\overline{x}$ and at a point x where $\dfrac{df}{dx} = 0$. When the function $f(x)$ is quadratic, its derivative is a linear function and therefore (unless we have a degenerate case) there is only one point where the derivative is equal to 0. So, to find the optimum of a quadratic function of one variable, it is sufficient to consider at most three values x (two if the point where the derivative is 0 lies outside the given interval).

For optimizing a function $f(x_1, \ldots, x_n)$ of several variables on the box $\underline{x}_i \leq x_i \leq \overline{x}_i$, the same trichotomy holds for each of the variables x_i: with respect to this variable, the optimum is attained either at one of the endpoints $\underline{x}_i$ and $\overline{x}_i$ and at a point x where the corresponding partial derivative is equal to 0 $\left(\dfrac{\partial f}{\partial x_i} = 0. \right)$

For each variable, we have only 3 options, but together, they form $3 \times \ldots \times 3 = 3^n$ options: e.g., when $x_1 = \underline{x}_1$, we still have 3 different options for x_2, etc. For each of these 3^n combinations of options, we have a system of linear equations to solve – which is relatively easy (see, e.g., [4]), but the shear amount of such cases makes this straightforward calculus-based algorithm exponential in time. The NP-hardness results proves, in effect, that unless P=NP, no other algorithm can solve this problem mush faster (in feasible polynomial time).

Ellipsoids: A Solution to the Computational Complexity Problem. One known solution to the above computational complexity problem is to use *ellipsoid* constraints instead of the boxes, i.e., to use constrains of the type $J(x_1, \ldots, x_n) \leq J_0$, where

$$J(x_1, \ldots, x_n) = b_0 + \sum_{i=1}^{n} b_i \cdot x_i + \sum_{i=1}^{n} \sum_{j=1}^{n} b_{ij} \cdot x_i \cdot x_j.$$

For such constraints, optimizing a quadratic function $f(x)$ means that:

– either the optimum is attained inside the ellipsoid, then we have a system
  of linear equations $\dfrac{\partial f}{\partial x_i} = 0$;
– or the optimum is attained on the border $J(x_1, \ldots, x_n) = J_0$ of the ellipsoid.

In the second case, the Lagrange multiplier approach leads to the unconstrained optimization of the auxiliary quadratic function $f + \lambda \cdot (J - J_0)$, i.e., again, to a solution of a system of linear equations. As a result, we get a solution $x(\lambda)$ as a function of λ.

The only additional problem is to find a single variable λ. This can be done in a relatively straightforward way, by solving an equation $J(x(\lambda)) = J_0$ with one unknown λ. The complexity of solving such an equation does not grow with the size n of the problem.

Computations can be made even more explicit if we take into account that if we have two quadratic forms, one of which is positive definite, we can move both to a diagonal form by applying an appropriate linear transformation; this linear transformation can be easily computed; see, e.g., [4]. Thus, when we apply an appropriate linear transformations of the coordinates, in the new coordinates $y_1, \ldots, y_n$, the ellipsoid $J \leq J_0$ becomes simply a unit circle $\sum_{i=1}^{n} y_i^2 = 1$, and the objective function takes the form

$$f(y_1, \ldots, y_n) = w_0 + \sum_{i=1}^{n} w_i \cdot y_i + \sum_{i=1}^{n} w_{ii} \cdot y_i^2.$$

In this case, the Lagrange functional takes the form

$$f_\lambda = w_0 + \sum_{i=1}^{n} w_i \cdot y_i + \sum_{i=1}^{n} w_{ii} \cdot y_i^2 + \lambda \cdot \left(\sum_{i=1}^{n} y_i^2 - 1 \right),$$

so equating partial derivatives of f_λ to 0 leads to $w_i + 2w_{ii} \cdot y_i + 2\lambda \cdot y_i = 0$, i.e., to $y_i = -\dfrac{w_i}{2 \cdot (w_{ii} + \lambda)}$, and the equation for λ takes the following explicit form:

$$\sum_{i=1}^{n} \frac{w_i^2}{4 \cdot (w_{ii} + \lambda)^2} = 1.$$

Because of this drastic reduction in computational complexity, ellipsoids have been successfully used in many applications; see, e.g., [1–3, 6, 7, 13, 14].

Need for Bisection. When the optimized function is simple – e.g., linear or quadratic – it does not matter how big or small is the area, the algorithm is the same. However, when the objective function is more complex, then for small areas, we can expand the objective function into Taylor series

$$f(x_1, \ldots, x_n) = a_0 + \sum_{i=1}^{n} a_i \cdot x_i + \sum_{i=1}^{n} \sum_{j=1}^{n} a_{ij} \cdot x_i \cdot x_j + \sum_{i=1}^{n} \sum_{j=1}^{n} \sum_{k=1}^{n} a_{ijk} \cdot x_i \cdot x_j \cdot x_k + \ldots,$$

and, with reasonable accuracy, keep only quadratic terms in this expansion.

For larger areas, such an approximation may not necessarily be sufficiently accurate. A similar problem occurs when we consider domains described by boxes. For boxes, a solution to this problem is straightforward: we divide ("bisect") the box into two sub-boxes by dividing one of the side intervals $[\underline{x}_i, \overline{x}_i]$ into two by a line $x_i = \widetilde{x}_i \stackrel{\text{def}}{=} \dfrac{\underline{x}_i + \overline{x}_i}{2}$. (In n-dimensional space, the equation $x_i = \widetilde{x}_i$ describes a plane.)

We can then try to estimate the optimum of the function over both sub-boxes, and, if necessary, further bisect each of these two sub-boxes into sub-sub-boxes [9, 12].

Bisection for Ellipsoids: A Problem. With ellipsoids, we can apply a similar idea: divide the ellipsoid into two halves by an appropriate plane. However, in comparison to boxes, here, we have an additional problem:

- when we bisect a box, both halves are boxes;
- however, a half of an ellipsoid is not an ellipsoid.

Thus, even when we know the algorithms for optimizing quadratic functions over ellipsoids, we cannot use them to optimize functions over half- or quarter-ellipsoids.

How This Problem Is Solved Now. At present (see, e.g., [2, 3, 10]), this problem is solved by enclosing each of the resulting half-ellipsoids into an ellipsoid. This procedure enables us to apply the same optimizations as before, but it comes with a price – that the enclosures are larger than the halves and thus, the size of the regions decreases slower that in the case of boxes – where, e.g., the volume of an area decreases by 2 on each bisection step. Since the areas do not go as fast, we will need more iterations (and thus, more computation time) to reach the desired small size.

Our Proposal. As an alternative, we propose to explicitly optimize quadratic functions over half-, quarter-, etc. ellipsoids.

Indeed, suppose that after a small number of bisections d, we have the resulting region. Each bisection j, $1 \leq j \leq d$, corresponds to selecting a half-space. Each half-space can be described by a linear inequality $\ell_j(x) \leq 0$, with a linear function $\ell_j(x)$. As before, for each j, the optimum is attained either inside the half-space or on its border, at a plane $\ell_j(x) \leq 0$. Thus, to find the desired optimum, we must check all 2^d subsets of the set $\{1, \ldots, d\}$. For each of these subsets S, we take all the planes $\ell_j(x) = 0$ with $j \in S$. The intersection of all these planes with the original ellipsoid is still an ellipsoid of smaller dimension. We then use the known ellipsoid-optimization algorithm to optimize the objective function over this smaller-dimension ellipsoid. The largest or smallest of the desired values is the desired maximum or minimum of the original objective function over our domain.

When d is small, the value 2^d is also small, so we still get an efficient algorithm.

Acknowledgments. This work was supported in part by NSF grants HRD-0734825 and DUE-0926721, and by Grant 1 T36 GM078000-01 from NIH.

References

1. Belforte, G., Bona, B.: An improved parameter identification algorithm for signal with unknown-but-bounded errors. In: Proceedings of the 7th IFAC Symposium on Identification and Parameter Estimation, NewYork, U.K. (1985)
2. Chernousko, F.L.: Estimation of the Phase Space of Dynamic Systems. Nauka Publ., Moscow (1988) (in Russian)
3. Chernousko, F.L.: State Estimation for Dynamic Systems. CRC Press, Boca Raton (1994)
4. Cormen, C.H., Leiserson, C.E., Rivest, R.L., Stein, C.: Introduction to Algorithms. MIT Press, Boston (2009)
5. Ferson, S., Ginzburg, L., Kreinovich, V., Longpré, L., Aviles, M.: Exact bounds on finite populations of interval data. Reliable Computing 11(3), 207–233 (2005)
6. Filippov, A.F.: Ellipsoidal estimates for a solution of a system of differential equations. Interval Computations 2(2(4)), 6–17 (1992)
7. Fogel, E., Huang, Y.F.: On the value of information in system identification. Bounded noise case. Automatica 18(2), 229–238 (1982)
8. Interval computations website, http://www.cs.utep.edu/interval-comp
9. Jaulin, L., Kieffer, M., Didrit, O., Walter, E.: Applied Interval Analysis, with Examples in Parameter and State Estimation, Robust Control and Robotics. Springer, London (2001)
10. Karmarkar, N.: A new polynomial-time algorithm for linear programming. Combinatorica 4, 373–396 (1984)
11. Kreinovich, V., Lakeyev, A., Rohn, J., Kahl, P.: Computational Complexity and Feasibility of Data Processing and Interval Computations. Kluwer, Dordrecht (1998)
12. Moore, R.E., Kearfott, R.B., Cloud, M.J.: Introduction to Interval Analysis. SIAM Press, Philadelphia (2009)
13. Schweppe, F.C.: Recursive state estimation: unknown but bounded errors and system inputs. IEEE Transactions on Automatic Control 13, 22 (1968)
14. Schweppe, F.C.: Uncertain Dynamic Systems. Prentice Hall, Englewood Cliffs (1973)
15. Vavasis, S.A.: Nonlinear Optimization: Complexity Issues. Oxford University Press, New York (1991)

An Auto-validating Rejection Sampler for Differentiable Arithmetical Expressions: Posterior Sampling of Phylogenetic Quartets

Raazesh Sainudiin

Laboratory for Mathematical Statistical Experiments and
Department of Mathematics and Statistics,
University of Canterbury,
Private Bag 4800, Christchurch, New Zealand 8041
r.sainudiin@math.canterbury.ac.nz
http://www.math.canterbury.ac.nz/~r.sainudiin

Abstract. We introduce an efficient extension of a recently introduced auto-validating rejection sampler that is capable of producing independent and identically distributed (IID) samples from a large class of target densities with locally Lipschitz arithmetical expressions. Our extension is restricted to target densities that are differentiable. We use the centered form, as opposed to the natural interval extension, to get tighter range enclosures of the differentiable multivariate target density using interval extended gradient differentiation arithmetic. By using the centered form we are able to sample one hundred times faster from the posterior density over the space of phylogenetic trees with four leaves (quartets).

Keywords: interval analysis, centered form, phylogentic inference.

1 Introduction

Obtaining independent and identical (IID) samples or realizations from a random vector T with probability density function $f(\cdot)$, denoted $T \sim f(\cdot)$, is a basic problem in computational statistics. The density $f(\cdot)(t) : \mathbb{T} \subseteq \mathbb{R}^d \to [0, \infty)$ allows one to obtain $\mathbb{P}(T \in B) = \int_B f(\cdot)(t)dt$, the probability that T belongs to any Borel set B. The density is absolutely continuous with respect to λ^d, the product of d Lebesgue measures, i.e., $f(\cdot) \ll \lambda^d$, and integrates to 1, i.e., $\int_{\mathbb{T}} f(\cdot)(t)dt = 1$. A *sampler* is a randomized algorithm that transforms independent and identically distributed (IID) samples from M, the uniformly distributed random variable on the unit interval, to those from the desired random object, say the random vector T with density $f(\cdot)$.

In Bayesian estimation, we want to draw IID samples from a target posterior density $f(\cdot)$ and more generally, in multivariate simulation, we want to draw IID samples from a random vector T with probability density $f(\cdot)$. These samples allow insights into the nature of the random vector itself. They are often used to estimate an integral of interest about the random vector, say, $\mathbb{E}_{f(\cdot)}(h(T)) :=$

M. Ceberio and V. Kreinovich (eds.), *Constraint Programming and Decision Making*,
Studies in Computational Intelligence 539,
DOI: 10.1007/978-3-319-04280-0_17, © Springer International Publishing Switzerland 2014

$\int_{\mathbb{T}} h(t) f(\cdot)(t) dt$, where $h(t) : \mathbb{T} \to \mathbb{R}$ is bounded and $\mathbb{E}_{f(\cdot)}(h^2(T)) < \infty$, using the estimator $\widehat{h}_n = n^{-1} \sum_{i=1}^{n} h(t_i)$, where $t_1, t_2, \ldots, t_n$ are IID samples from T with density $f(\cdot)$. For example, such integrals of interest can be the posterior mean given by $\int_{\mathbb{T}} t f(\cdot)(t) dt$ with $h(t) = t$, or the probability of an event A given by $\mathbb{P}(A) := \int_{\mathbb{T}} \mathbb{1}_A(t) f(\cdot)(t) dt$, with $h(t) = \mathbb{1}_A(t)$, where $\mathbb{1}_A(t)$ equals 1 if $t \in A$ and 0 otherwise. Due to the strong law of large numbers our estimator $\widehat{h}_n$ converges to the desired $\mathbb{E}_{f(\cdot)}(h(T))$ with probability 1 as the number of samples n approaches infinity. Furthermore, the condition $\mathbb{E}_{f(\cdot)}(h^2(T)) < \infty$ ensures the asymptotic normality of our estimator due to the central limit theorem and provides a straightforward calculation of a confidence interval.

In Bayesian phylogenetic estimation, we want to draw independent and identically distributed samples from a target posterior density on the space of phylogenetic trees. The standard approaches to sampling from the posterior density, especially over phylogenetic trees, rely on Markov chain Monte Carlo (MCMC) methods. Despite their asymptotic validity, it is nontrivial to guarantee that an MCMC algorithm has converged to stationarity [5], and thus MCMC convergence diagnostics on phylogenetic tree spaces are heuristic and may lead to meaningless estimates [8].

A more direct method for simulating IID samples from a random variable T with density $f(\cdot)(t)$ is the *Rejection Sampler* (RS) of von Neumann [15]. RS can produce IID samples from the target density $f(\cdot)(t) := f(t)/(N_f)$ by only evaluating the *target shape* $f(t)$ — without knowing the normalising constant $N_f := \int_{\mathbb{T}} f(t) dt$. Briefly, the idea behind RS is as follows: produce a point uniformly distributed in the $(d+1)$-dimensional region under an envelope function that is strictly greater than or equal to the target shape and if this point is below the target shape then accept its first d coordinates in $\mathbb{T}$ as a sample from T, otherwise reject it and try again.

RS can produce samples from $T \sim f(\cdot)$ according to Algorithm 1 when provided with (i) a fundamental sampler that can produce independent samples from the Uniform$[0,1]$ random variable M with density $\mathbb{1}_{[0,1]}(m) : \mathbb{R} \mapsto \mathbb{R}$, (ii) a target shape $f(t) : \mathbb{T} \mapsto \mathbb{R}$, (iii) an envelope function $\widehat{g}(t) : \mathbb{T} \mapsto \mathbb{R}$, such that,

$$\widehat{g}(t) \geq f(t) \quad \text{for all } t \in \mathbb{T} , \tag{1}$$

(iv) a normalizing constant $N_{\widehat{g}} := \int_{\mathbb{T}} \widehat{g}(t) dt$, (v) a proposal density $g(t) := (N_{\widehat{g}})^{-1} \widehat{g}(t)$ over $\mathbb{T}$ from which independent samples can be drawn and finally (vi) $f(t)$ and $\widehat{g}(t)$ must be computable for any $t \in \mathbb{T}$.

The random variable T, if generated by Algorithm 1, is distributed according to $f(\cdot)$ (e.g. [17]). Let $\boldsymbol{A}(\widehat{g})$ be the probability that a point proposed according to g gets accepted as an independent sample from $f(\cdot)$ through the envelope function $\widehat{g}$. Observe that the envelope-specific acceptance probability $\boldsymbol{A}(\widehat{g})$ is the ratio of the integrals

$$\boldsymbol{A}(\widehat{g}) = \frac{N_f}{N_{\widehat{g}}} := \frac{\int_{\mathbb{T}} f(t) \, dt}{\int_{\mathbb{T}} \widehat{g}(t) \, dt} ,$$

Algorithm 1. von Neumann RS

  input   : (i) f; (ii) samplers for $V \sim g$ and $M \sim \mathbf{1}_{[0,1]}$; (iii) $\widehat{g}$; (iv) integer MaxTrials;
  output   : (i) possibly one sample t from $T \sim f(\cdot)$ and (ii) Trials

  initialize: Trials $\leftarrow$ 0; Success $\leftarrow$ false; $t \leftarrow \emptyset$;

  repeat   // propose at most MaxTrials times until acceptance
  $\quad$ | $v \leftarrow$ **sample**(g) ;   // draw a sample v from RV V with density g
  $\quad$ | $u \leftarrow \widehat{g}(v)$ **sample**$(\mathbf{1}_{[0,1]})$;   // draw a sample u from RV U with density $\mathbf{1}_{[0,\widehat{g}(v)]}$
  $\quad$ | **if** $u \leq f(v)$ **then**   // accept the proposed v and flag Success
  $\quad$ | $\quad$ | $t \leftarrow v$; Success $\leftarrow$ true
  $\quad$ | **end**
  $\quad$ | Trials $\leftarrow$ Trials $+1$;   // track the number of proposal trials so far
  until Trials $\geq$ MaxTrials *or* Success $=$ true;
  return t *and* Trials

---

and the probability distribution over the number of samples from g to obtain one sample from $f(\cdot)$ is geometrically distributed with mean $1/\boldsymbol{A}(\widehat{g})$ (e.g. [17]).

The crucial step in RS is the construction of an envelope function $\widehat{g}(t)$ that is not only greater than the target shape $f(t) := N_f f(\cdot)(t)$ at every $t \in \mathbb{T} \subseteq \mathbb{R}^d$, but also easy to normalise and draw samples from. Moreover, a practical and efficient envelope function has to be as close to the target shape as possible from above in order to ensure a sufficiently high acceptance probability.

Moore rejection sampler (MRS) of [11, 12] uses the natural interval extension of f over an adaptive partition of $\mathbb{T}$ to rigorously produce IID samples from the posterior distribution over phylogenetic tree spaces. Informally, MRS [11, 12] partitions the domain into boxes and uses interval analysis to rigorously enclose the range of the target shape in each box; then it uses as envelope the piece-wise constant function given by the upper bound of the range in each box. More formally, the method employs the natural interval extension of the target posterior shape $f(t) : \mathbb{T} \mapsto \mathbb{R}$ to produce rigorous enclosures of the range of f over each interval vector or box in an adaptive partition $\mathfrak{T} := \{t^{(1)}, t^{(2)}, \ldots, t^{(|\mathfrak{T}|)}\}$ of the tree space $\mathbb{T} = \cup_i t^{(i)}$. This partition is adaptively constructed by a priority queue. The interval extended target shape maps boxes in $\mathbb{T}$ to intervals in $\mathbb{R}$. This image interval provides an upper bound for the global maximum and a lower bound for the global minimum of f over each element of the partition of $\mathbb{T}$. This information is used to construct an envelope as a simple function over the partition $\mathfrak{T}$. Using the Alias method [16] samples are proposed efficiently from this normalized piece-wise constant function envelope for von Neumann rejection sampling. Unlike many conventional samplers, each sample produced by MRS is equivalent to a computer-assisted proof that it is drawn from the desired target density.

2   An Improved Approach

Suppose our target shape $f : \mathbb{T} \to \mathbb{R}$, with $\mathbb{T} \subseteq \mathbb{R}^d$, is differentiable. Then the centered form [7, 10] (of first order) of f over a box $\boldsymbol{x} \in \mathbb{T}$ is

$$\boldsymbol{f}_c(\boldsymbol{x}) := (f(c) + [\nabla f(\boldsymbol{x})](\boldsymbol{x} - c)) \cap \boldsymbol{f}(\boldsymbol{x})$$

where, $c := \text{mid}\,(\boldsymbol{x})$ is the midpoint of the box $\boldsymbol{x}$, $\boldsymbol{f}(\boldsymbol{x})$ is the natural interval extension of f over $\boldsymbol{x}$ and $[\nabla f(\boldsymbol{x})]$ is the enclosure of the gradient of f over $\boldsymbol{x}$. For any box $\boldsymbol{x} \subseteq \mathbb{T}$ it is well known that the centered form encloses the range, i.e.,

$$f(\boldsymbol{x}) := \{f(x) : x \in \boldsymbol{x}\} \subseteq \boldsymbol{f}_c(\boldsymbol{x}) \subseteq \boldsymbol{f}(\boldsymbol{x}) \ ,$$

and the range enclosure given by $\boldsymbol{f}_c(\boldsymbol{x})$ is usually sharper than $\boldsymbol{f}(\boldsymbol{x})$ especially as $\boldsymbol{x}$ shrinks. Instead of using the natural interval extension $\boldsymbol{f}(\boldsymbol{x})$ to enclose $f(\boldsymbol{x})$, the range of f over $\boldsymbol{x}$, as done in [11, 12], our new approach employs the centered form $\boldsymbol{f}_c(\boldsymbol{x})$ to enclose the range of the target shape over each box $\boldsymbol{x}$ in the partition of $\mathbb{T}$.

We use gradient differentiation arithmetic using interval-extended automatic differentiation (e.g. [9]) to obtain $[\nabla f(\boldsymbol{x})]$, the enclosure of the gradient of f over $\boldsymbol{x}$. We use the implementation of this arithmetic in the grad_ari module of C-XSC 2.0, a C++ class library for extended scientific computing [6]. Using the centered form to bound the range in MRS allows us to sample from the posterior density over the space of phylogenetic quartets 100 times faster than using the natural interval extension alone as in [11]. An open source C++ class library for MRS is available from www.math.canterbury.ac.nz/~r.sainudiin/codes/mrs under the terms of the GNU general public license (GPL).

3   Phylogenetic Estimation

In this section we briefly review phylogenetic estimation as we will apply our improved method to sample from challenging phylogenetic posterior densities. Introduction to phylogenetics can be found in [14, 19]. Inferring the ancestral relationship among a set of extant (presently surviving) species based on their DNA sequences is a basic problem in phylogenetic estimation. A phylogenetic tree relates the extant species represented by its leaf nodes with ancestral species represented by its internal nodes. The topology or shape of the tree specifies the order of speciation or branching events. The length of an edge (branch length) connecting two nodes (species) in the phylogenetic tree represents the amount of evolutionary time (divergence) between the two species as measured by the differences in their DNA sequence due to mutation. One can obtain the likelihood of a particular phylogenetic tree that relates the extant species of interest at its leaves by superimposing a continuous time Markov chain model of DNA mutation along the lengths of the branches on that tree. During the likelihood computation, one needs to sum over all possible states of the DNA sequence at the unobserved ancestral nodes. In [11] MRS was used to draw IID posterior samples from small phylogenetic tree spaces of the same dimension (number of branches) based on primate DNA sequence data. In [12] this was generalized to the trans-dimensional setting where the number of branch length parameters are allowed to vary between models of phylogenetic trees. However, only the natural interval extension of the posterior density was used in [11] and [12] to obtain upper bounds for the range before Moore rejection sampling. Here, we will employ interval extended gradient differentiation arithmetic to obtain much tighter enclosures of the posterior density, which is the product of a uniform prior density and the likelihood function over phylogenetic trees.

Likelihood of a Phylogenetic Tree. Let d denote a homologous set of sequences of length v with character set $\mathfrak{A} = \{a_1, a_2, \ldots, a_{|\mathfrak{A}|}\}$ from n taxa. We think of d as an $n \times v$ matrix with entries from $\mathfrak{A}$. We are interested in estimating the branch lengths and topologies of the tree underlying our observed d. Let b_k denote the number of branches and s_k denote the number of nodes of a tree with a specific topology or branching order labeled by k. Thus, for a given topology label k, n labeled leaves and b_k many branches, the labeled tree $^k t$ is the topology-labeled vector of branch lengths $(^k t_1, \ldots, ^k t_{b_k})$ contained in the topology-labeled tree space $^k \mathbb{T}$, i.e.,

$$^k \mathbb{T} := \{^k t := (^k t_1, \ldots, ^k t_{b_k}) \in \mathbb{R}_+^{b_k} : {}^k t_i > 0 \text{ for terminal branches}\} \ .$$

The tree space with $|\mathfrak{K}|$ many topologies in the topology label set $\mathfrak{K}$ can be defined as follows:

$$^{\mathfrak{K}} \mathbb{T} := \bigcup_{k \in \mathfrak{K}} {}^k \mathbb{T} \ .$$

An explicit model of sequence evolution is prescribed in order to obtain the likelihood of observing data d at the leaf nodes as a function of the parameter $^k t \in {}^{\mathfrak{K}} \mathbb{T}$ for each topology label $k \in \mathfrak{K}$. Such a model prescribes $P_{a_i, a_j}(t)$, the probability of mutation from a character $a_i \in \mathfrak{A}$ to another character $a_j \in \mathfrak{A}$ in time t. Using such a transition probability we may compute $\ell_q(^k t)$, the log-likelihood of the data d at site $q \in \{1, \ldots, v\}$ or the q-th column of d, via the post-order traversal over the labeled tree with branch lengths $^k t := (^k t_1, {}^k t_2, \ldots, {}^k t_{b_k})$. This amounts to the sum-product Algorithm 2 [3] that associates with each node $h \in \{1, \ldots, s_k\}$ of $^k t$ subtending $\hbar$ many descendants, a partial likelihood vector, $l_h := (l_h^{(a_1)}, l_h^{(a_2)}, \ldots, l_h^{(a_{|\mathfrak{A}|})}) \in \mathbb{R}^{|\mathfrak{A}|}$, and specifies the length of the branch leading to its ancestor as $^k t_h$.

Assuming independence across all v sites we obtain the likelihood function for the given data d, by multiplying the site-specific likelihoods

$$l_d(^k t) = \prod_{q=1}^{v} l_{d_{\cdot,q}}(^k t) \ . \tag{2}$$

The maximum likelihood estimate is a point estimate (single best guess) of the unknown phylogenetic tree on the basis of the observed data d and it is

$$\operatorname*{argmax}_{^k t \in {}^{\mathfrak{K}} \mathbb{T}} l_d(^k t) \ .$$

Algorithm 2. Likelihood by post-order traversal

input   : (i) a labeled tree with branch lengths ${}^k t := ({}^k t_1, {}^k t_2, \ldots, {}^k t_{b_k})$, (ii)
transition probability $P_{a_i, a_j}(t)$ for any $a_i, a_j \in \mathfrak{A}$, (iii) stationary
distribution $\pi(a_i)$ over each character $a_i \in \mathfrak{A}$, (iv) site pattern or data
$d_{.,q}$ at site q

output   : $l_{d_{.,q}}({}^k t)$, the likelihood at site q with pattern $d_{.,q}$

initialize: For a leaf node h with observed character $a_i = d_{h,q}$ at site q, set
$l_h^{(a_i)} = 1$ and $l_h^{(a_j)} = 0$ for all $j \neq i$. For any internal node h, set
$l_h := (1, 1, \ldots, 1)$.

recurse : compute l_h for each sub-terminal node h, then those of their
ancestors recursively to finally compute l_r for the root node r to obtain
the likelihood for site q,

$$l_{d_{.,q}}({}^k t) = l_r = \sum_{a_i \in \mathfrak{A}} \left(\pi(a_i) \cdot l_r^{(a_i)} \right) .$$

For an internal node h with descendants $s_1, s_2, \ldots, s_\hbar$,

$$l_h^{(a_i)} = \sum_{j_1, \ldots, j_\hbar \in \mathfrak{A}} \left\{ l_{s_1}^{(j_1)} \cdot P_{a_i, j_1}({}^k t_{s_1}) \cdot l_{s_2}^{(j_2)} \cdot P_{a_i, j_2}({}^k t_{s_2}) \ldots l_{s_\hbar}^{(j_\hbar)} \cdot P_{a_i, j_\hbar}({}^k t_{s_\hbar}) \right\}.$$

The simplest probability models for character mutation are continuous time
Markov chains with finite state space $\mathfrak{A}$. We introduce the simplest such model
with just two characters as it is thought to well-represent the core problems in
phylogenetic estimation (see for e.g. [18]).

Posterior Density of a Tree. The posterior density $f(\cdot)({}^k t)$ conditional on
data d at tree ${}^k t$ is the normalized product of the likelihood $l_d({}^k t)$ and the prior
density $p({}^k t)$ over a given tree space ${}^{\mathfrak{K}}\mathbb{T}$:

$$f(\cdot)({}^k t) = \frac{l_d({}^k t) p({}^k t)}{\int_{\mathfrak{K}\mathbb{T}} l_d({}^k t) p({}^k t) \, \partial({}^k t)} . \tag{3}$$

We assume a uniform prior density over a large box or a union of large boxes in
a given tree space ${}^{\mathfrak{K}}\mathbb{T}$. Typically, the sides of the box giving the range of branch
lengths, are extremely long, say, $[0, 10]$ or $[10^{-10}, 10]$. The branch lengths are
measured in units of expected number of DNA substitutions per site and there-
fore the support of our uniform prior density over ${}^{\mathfrak{K}}\mathbb{T}$ contains the biologically
relevant branch lengths. If ${}^{\mathfrak{K}}\mathbb{T}$ is a union of distinct topologies then we let our
prior be an equally weighted finite mixture of uniform densities over large boxes
in each topology. Naturally, other prior densities are possible especially in the
presence of additional information. We choose flat priors for the convenient in-
terpretation of the target posterior shape $f({}^k t) = f(\cdot)({}^k t) \int_{\mathfrak{K}\mathbb{T}} l_d({}^k t) p({}^k t) \, \partial({}^k t)$ to

be the likelihood function in the absence of prior information beyond a compact support specification.

Likelihood under Cavender-Farris-Neyman (CFN) Model. We now describe the simplest model for the evolution of binary sequences under a symmetric transition matrix over all branches of a tree. This model has been used by authors in various fields including molecular biology, information theory, operations research and statistical physics; for references see [2, 14]. This model is referred to as the Cavender-Farris-Neyman (CFN) model in molecular biology, although in other fields it has been referred to as 'the on-off machine', 'symmetric binary channel' and the 'symmetric two-state Poisson model'.

Under the CFN mutation model, only pyrimidines and purines, denoted respectively by $Y := \{C, T\}$ and $R := \{A, G\}$, are distinguished as evolutionary states among the four nucleotides $\{A, G, C, T\}$, i.e. $\mathfrak{A} = \{Y, R\}$. Time t is measured by the expected number of substitutions in this homogeneous continuous time Markov chain with rate matrix:

$$Q = \begin{pmatrix} -1 & 1 \\ 1 & -1 \end{pmatrix} \ ,$$

and transition probability matrix $P(t) = e^{Qt}$:

$$P(t) = \begin{pmatrix} 1 - (1 - e^{-2t})/2 & (1 - e^{-2t})/2 \\ (1 - e^{-2t})/2 & 1 - (1 - e^{-2t})/2 \end{pmatrix} \ .$$

Thus, the probability that Y mutates to R, or vice versa, in time t is $a(t) := (1 - e^{-2t})/2$. The stationary distribution is uniform on $\mathfrak{A}$, i.e. $\pi(R) = \pi(Y) = 1/2$.

To get a concerete idea of the likelihood function let us consider the case when there are only three taxa. Consider the unrooted tree space with a single topology labeled 0 and three non-negative terminal branch lengths ${}^0t = ({}^0t_1, {}^0t_2, {}^0t_3) \in \mathbb{R}_+^3$ as shown in Figure 1 (i). There are $2^3 = 8$ possible site patterns, i.e. for each site $q \in \{1, 2, \ldots, v\}$, the q-th column of the data d, denoted by $d_{.,q}$, is one of eight possibilities, numbered $0, 1, \ldots, 7$ for convenience:

$$d_{.,q} \in \begin{Bmatrix} 0 \ , 1 \ , 2 \ , 3 \ , 4 \ , 5 \ , 6 \ , 7 \\ R \ \ Y \ \ R \ \ Y \ \ R \ \ Y \ \ R \ \ Y \\ R \ , Y \ , R \ , Y \ , Y \ , R \ , Y \ , R \\ R \ \ Y \ \ Y \ \ R \ \ Y \ \ R \ \ R \ \ Y \end{Bmatrix} \ . \tag{4}$$

Given a multiple sequence alignment data d from 3 taxa at v homologous sites, i.e. $d \in \{Y, R\}^{3 \times v}$, Algorithm 2 can be used to compute the likelihood of any tree ${}^0t \in {}^0\mathbb{T}$ as follows:

$$l_d({}^kt) = \prod_{q=1}^{v} l_{d_{.,q}}({}^kt) = \prod_{i=0}^{7} \left(l_i({}^kt) \right)^{c_i} \ , \tag{5}$$

where $l_i({}^kt)$ is the likelihood of the the i-th site pattern as in (4) and c_i is the count of sites with pattern i. In fact, $l_i({}^kt) = P(i|{}^kt)$ is the probability of observing site pattern i given topology label k and branch lengths t and similarly $l_d(^kt) = P(d|^kt)$. By using gradient differentiation arithmetic in Algorithm 2 we can obtain the enclosure of $l_d(^kt)$, the likelihood of a box ^kt in the tree space $^k\mathbb{T}$, using the centered form. When there are four taxa we have three unrooted topologies, each with five branch length parameters, as shown in Figure 1 (ii), (iii) and (iv). In this case there are sixteen site patterns and we can similarly obtain the likelihood for each topology using Algorithm 2.

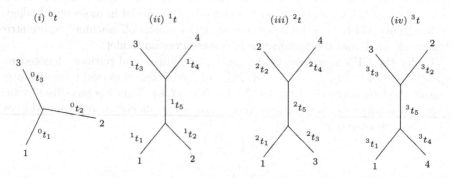

Fig. 1. The only topology of the unrooted tree with three taxa is shown in (i) and the three unrooted trees on four taxa with five branch length parameters each are shown by (ii), (iii) and (iv), respectively

4 Efficiency of MRS with Centered Form

For typical data sets on three taxa, including those in [12, Table 3], posterior samples of size 10^5 using a partition of size 10^3 can be obtained about 100 times faster by using the centered form as opposed to the natural interval extension of the likelihood function. This speed-up is particularly significant for typical data sets on four taxa as posterior samples of size 10^6 are available in about $30 - 150$ minutes as opposed to a few days.

We were able to produce 10^7 posterior samples in under two hours for all of the synthetic data sets in [1]. These data sets are well-known in phylogenetics for producing multiple maxima. The matrix plot of the first 10^4 posterior samples in tree space $^1\mathbb{T}$ is shown in Figure 2 for the following site pattern counts from [1, Proof of Thm. 2]:

$$c_{\text{YYYY}} = 1400, \quad c_{\text{RYYY}} = 1, \quad c_{\text{RYRR}} = 1, \quad c_{\text{RRYY}} = 300,$$
$$c_{\text{RRYR}} = 1, \quad c_{\text{RYRY}} = 200, \quad c_{\text{YRRY}} = 100, \quad c_{\text{RRRY}} = 1 \ . \quad (6)$$

The likelihood function for this data is known to attain the maximum at two distinct points in $^1\mathbb{T}$. This is evident from the two clusters of posterior samples in the matrix plot of Figure 2.

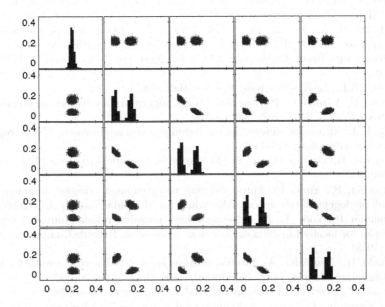

Fig. 2. The matrix plot of 10^4 posterior samples for the site pattern counts in (6)

5 Conclusion

In this paper, we use the centered form of the posterior density using gradient differentiation arithmetic to get tighter range enclosures and thereby increase the acceptance probability of the naive Moore rejection sampler [11, 12] that is only based on the natural interval extension of the posterior density. When we apply this centered form to produce IID samples from phylogenetic posterior densities over three and four taxa tree spaces we observe a hundred-fold speedup. Higher-order centered forms [10] in conjunction with constraint propagation [13] may further improve the sampler efficiency enough to produce IID posterior samples from five taxa phylogenetic trees with fifteen topologies in seven dimensions.

References

1. Chor, B., Hendy, M., Holland, B., Penny, D.: Multiple maxima of likelihood in phylogenetic trees: An analytic approach. Mol. Biol. Evol. 17, 1529–1541 (2000)
2. Evans, W., Kenyon, C., Peres, Y., Schulman, L.J.: Broadcasting on trees and the Ising model. Advances in Applied Probability 10, 410–433 (2000)
3. Felsenstein, J.: Evolutionary trees from DNA sequences: a maximum likelihood approach. J. Mol. Evol. 17, 368–376 (1981)
4. Felsenstein, J.: Inferring phylogenies. Sinauer Associates, Sunderland (2003)

5. Jones, G., Hobert, J.: Honest exploration of intractable probability distributions via Markov chain Monte Carlo. Statistical Science 16(4), 312–334 (2001)
6. Hofschuster, W., Krämer, W.: C-XSC 2.0 – A C++ library for extended scientific computing. In: Alt, R., Frommer, A., Kearfott, R.B., Luther, W. (eds.) Num. Software with Result Verification. LNCS, vol. 2991, pp. 15–35. Springer, Heidelberg (2004)
7. Moore, R.E.: Interval analysis. Prentice-Hall (1967)
8. Mossel, E., Vigoda, E.: Phylogenetic MCMC algorithms are misleading on mixtures of trees. Science 309, 2207–2209 (2005)
9. Rall, L.B.: Automatic differentiation, techniques and applications. LNCS, vol. 120. Springer, Heidelberg (1981)
10. Ratschek, H.: Centered forms. SIAM Journal on Numerical Analysis 17(5), 656–662 (1980)
11. Sainudiin, R., York, T.: Auto-validating von Neumann rejection sampling from small phylogenetic tree spaces. Algorithms for Molecular Biology 4, 1 (2009)
12. Sainudiin, R., York, T.: An auto-validating, trans-dimensional, universal rejection sampler for locally Lipschitz arithmetical expressions. Reliable Computing 18, 15–54 (2013)
13. Schichl, H., Neumaier, A.: Interval analysis on directed acyclic graphs for global optimization. Journal of Global Optimization 33(4), 541–562 (2005)
14. Semple, C., Steel, M.: Phylogenetics. Oxford University Press (2003)
15. von Neumann, J.: Various techniques used in connection with random digits. In: Von Neumann, J. (ed.) Collected Works, vol. V. Oxford University Press (1963)
16. Walker, A.J.: An efficient method for generating discrete random variables with general distributions. ACM Trans. Math. Softw. 3, 253–256 (1977)
17. Williams, D.: Weighing the Odds: A Course in Probability and Statistics. Cambridge University Press (2001)
18. Yang, Z.: Complexity of the simplest phylogenetic estimation problem. Proceedings of the Royal Soc. London B Biol. Sci. 267, 109–119 (2000)
19. Yang, Z.: Computational Molecular Evolution. Oxford University Press, UK (2006)

Graph Subdivision Methods in Interval Global Optimization

Sergey P. Shary

Institute of Computational Technologies,
Novosibirsk, Russia

Abstract. The work advances a new class of global optimization methods, called *graph subdivision methods*, that are based on simultaneous adaptive subdivision of both the function's domain of definition and the range of values. An application to interval linear systems is given.

Keywords: global optimization, interval analysis, adaptive subdivision.

1 Introduction

The subject matter of our paper is the problem of global optimization of a real-valued function $f : \mathbb{R}^n \supseteq X \to \mathbb{R}$ over an axis-aligned rectangular box X (i. e. over an interval vector):

$$\text{find} \quad \min_{x \in X} f(x). \tag{1}$$

The problem (1) is known to be (more or less) successfully solved by various interval techniques [1, 3, 6], which enables one to reliably compute two-sided bounds for both the optimum value and the argument it is attained at. The basis of these methods is adaptive, according to the "branch-and-bound" strategy, subdivision of the domain of the function to be minimized combined with the interval evaluation of the ranges over the resulting subdomains.

The purpose of our work is to present a new promising interval approach for the solution of the problem (1) that relies upon joint adaptive subdivision of both the function's domain of definition and its range of values. For some classes of problems, the new approach is expected to turn out better than the traditional techniques from [1, 3, 6] in either implementation flexibility or computational efficacy and the quality of the results it produces. A shortened version of this article has been previously published as [9].

2 Idea of the New Approach

Notice that any function $f : \mathbb{R}^n \supseteq X \to \mathbb{R}$, being by the very definition a special subset of the direct product $\mathbb{R}^n \times \mathbb{R}$, is an $(n+1)$-dimensional object. In connection with it, we usually use the concept of the *graph* of the function f:

$$\text{graph of } f = \left\{ (x, t) \in \mathbb{R}^{n+1} \mid x \in \mathbb{R}^n, t \in \mathbb{R}, f(x) = t \right\}.$$

M. Ceberio and V. Kreinovich (eds.), *Constraint Programming and Decision Making*,
Studies in Computational Intelligence 539,
DOI: 10.1007/978-3-319-04280-0_18, © Springer International Publishing Switzerland 2014

However, the interval global optimization methods that we mentioned in Introduction involve into active operation — adaptive subdivision — only the first n coordinates of this set. The last $(n + 1)$-th coordinate of the function represented by its graph is processed in a substantially different manner, passively, and the same is true for the overwhelming majority of the classical optimization techniques. How could we correct the situation and what would be the result?

We start our consideration from the simplest case of a single-variable function $f : \mathbb{R} \supseteq X \to \mathbb{R}$, defined on a closed interval X, for which we have to solve the problem (1). In the plane $0xy$, let us construct a straight line parallel to the first axis, with the equation $y = l$, where l is a constant. We can ascertain whether the line intersects the graph of the function $y = f(x)$ after having solved the equation

$$f(x) - l = 0 \tag{2}$$

on X or, alternatively, making sure that it is incompatible (unsolvable). As is easily seen, the answer to the above question provides us with information on the minimum (1) under computation: if the straight line $y = l$ intersects the graph of the function $y = f(x)$, then

$$\min_{x \in X} f(x) \le l.$$

Moreover, if $f(x)$ is continuous on X, then

$$\min_{x \in X} f(x) = \min\{ l \in \mathbb{R} \mid \text{ the equation } f(x) - l = 0 \text{ is solvable} \}.$$

Therefore, varying the value of the "level" l and repeating the process of the solution of the equation (2), we can substantially improve the estimate for the sought-for minimum (1).

The procedure we have just described can be substantially modified by using the ideas and methods of the interval analysis:

First, the interval methods make it possible to easily compute estimates for the range of f over X from below and from above, which is necessary to determine the bounds of variation of the level l in the process of the correction of the minimum.

Second, it makes sense to examine the intersection of the graph of the function $y = f(x)$ not with single lines, but with the whole bundles of lines parallel to the $0x$ axis and defined by the equations $y = \boldsymbol{l}$, where $\boldsymbol{l}$ is an interval in $\mathbb{R}$. We will be able thereby to estimate the global minimum (1) both from below and from above, since $\min_{x \in X} f(x)$ is not less than the minimum of the left endpoints and not greater than the minimum of the right endpoints of all the intervals $\boldsymbol{l}$ such that the bundle $y = \boldsymbol{l}$ intersects the graph of the function $y = f(x)$.

Third, the interval methods for the solution of equations (e.g. the interval Newton method and its modifications [1, 3, 5]) enable us, under very mild requirement on the smoothness of f, to examine solvability of both the point equation (2) and the interval equation $f(x) - \boldsymbol{l} = 0$. The latter is understood as the existence of some $l \in \boldsymbol{l}$ for which (2) is solvable.

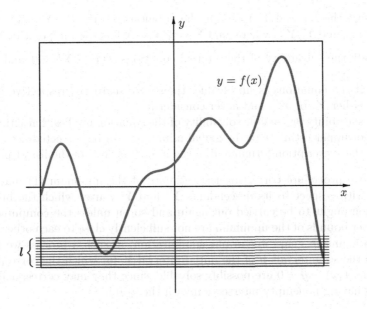

Fig. 1. Does the bundle of lines intersect the graph of the function?...

The answer produced by the interval methods may have one of the following forms [1, 3, 5]:

1. The equation does not have solutions — *unsolvable* — within the interval under consideration, i. e. $0 \notin f(x) - l$ for any $x \in \boldsymbol{X}$.

2. The equation has, with guarantee, a solution (or solutions) within the interval under consideration, i. e. there certainly exists $x^\star \in \boldsymbol{X}$ such that $f(x^\star) - l = 0$ for at least one $l \in \boldsymbol{l}$. We shall speak that the equation is just *solvable* then.

3. Applying the solution procedure does not allow us to speak, with certainty, that the equation is either has solutions or unsolvable on the given interval $\boldsymbol{X}$. In such cases, we shall speak that the equation is *possibly solvable*.

The third option is the most unfavorable algorithmically, but we should carefully take it into account in our reasoning since this kind of uncertainty is quite actual in computation often being the case when the equation (2) has multiple roots. Notice also that the interval methods never lose roots and cannot at all output the message "no solutions" if the equation really has them.

Finally, we will use the subdivision of the interval of the range of values instead of the "varying the level" l. The overall interval version of the procedure for finding the global minimum of the single-variable function $f(x)$ over the interval $\boldsymbol{X}$ can look as follows. It starts with computing a crude interval enclosure $\boldsymbol{Y}$ of the range of values $f(x)$ over $\boldsymbol{X}$ (for example, as the natural interval extension of f on $\boldsymbol{X}$). Further, we bisect the interval $\boldsymbol{Y}$ to beget the subintervals $\boldsymbol{Y}' := [\underline{\boldsymbol{Y}}', \text{mid } \boldsymbol{Y}]$ and $\boldsymbol{Y}'' := [\text{mid } \boldsymbol{Y}, \overline{\boldsymbol{Y}}]$, where $\text{mid } \boldsymbol{Y} = \frac{1}{2}(\overline{\boldsymbol{Y}} + \underline{\boldsymbol{Y}})$ is the midpoint of $\boldsymbol{Y}$;

we check the solvability of the interval equations $f(x) - \boldsymbol{Y}' = 0$ and $f(x) - \boldsymbol{Y}'' = 0$:

- if the equations is unsolvable, then we discard the respective interval, either $\boldsymbol{Y}'$ or $\boldsymbol{Y}''$, and never consider it;
- solvability or possible solvability of the equation implies that either lower or upper estimate of the global minimum can be corrected according to the prescription formulated in the item "Second" of the list at page 154.

The above procedure correcting the estimate of the minimum (1) may be repeated with respect to its descendants $\boldsymbol{Y}'$ and $\boldsymbol{Y}''$, after which the bisection-correction ought to be carried out again and so on unless the computed lower and upper bounds of the minimum are not sufficiently close to each other. Notice that, to maintain guarantee of our computation, in such a process we have to keep all the subintervals $\boldsymbol{y}$ of the initial interval $\boldsymbol{Y}$ for which the corresponding equations $f(x) - \boldsymbol{y} = 0$ are possibly solvable, since they may correspond to the bundles having nonempty intersection with the graph.

3 Multidimensional Case

Theoretically, the computational scheme of the one-dimensional global optimization algorithm we have developed in the preceding section is completely applicable to the functions $f(x) := f(x_1, x_2, \ldots, x_n)$ of several variables. The only thing we should be able to do for that is to check intersection of the graph of the function $y = f(x)$ with the bundle of the hyperplanes $y = l$ that are orthogonal to the $0y$ axis. Sometimes, that can be really done when we have a powerful equations solver and are able to apply it easily. In particular, Semenov [7] implemented a similar kind of procedure to refine the value of the optimum in some problems.

However, in most cases the practical implementation of our idea encounters big difficulties. The point is that, in the general multidimensional case, the solution of an equation — inquiring into its solvability — is in no ways easier problem than the global optimization. As opposed to the single-variable situation, we do not have simple and efficient techniques such as the interval Newton method and its modifications at our disposal. A way out of the difficulty may be subdivision of the domain of definition of f — the box $\boldsymbol{X}$ — along some (but not all!) selected coordinate directions, whose number and specific choice depend on the problem under solution and its objective function.

The coordinate directions along which the function's domain shall not be subdivided will be referred to as *mute*, and first we consider the simplest methods having only one mute direction with the number $\mu \in \{1, 2, \ldots, n\}$. Let, in the

Table 1. The simplest graph subdivision method for global optimization * (one mute variable)

Input

A box $X \subseteq \mathbb{R}^n$ and a function $f : X \to \mathbb{R}$.   An accuracy $\epsilon > 0$.

A number μ of the mute component, $\mu \in \{1, 2, \ldots, n\}$.

A method for checking the solvability of the single-variable interval equation $\phi(Z, t) = 0$ for ϕ and Z defined as (4)–(5).

Output

The lower $\underline{y}$ and upper $\overline{y}$ estimates, with the accuracy ϵ, for the global minimum of the function f over the box X .

Algorithm

compute an enclosure Y of the range of f over X ;

assign $Z := (X_1, \ldots, X_{\mu-1}, X_{\mu+1}, \ldots, X_n, Y)$;

set $z := \underline{Y}$ and $\overline{y} := \overline{Y}$;

initialize the working list $\mathcal{L} := \big\{ (Z, z) \big\}$;

DO WHILE $(\overline{y} - z \geq \epsilon)$

    choose the component k of the box Z having the largest length, i. e. such that wid $Z_k = \max_{1 \leq i \leq n}$ wid Z_i ;

    bisect the box Z along the k-th coordinate to get the boxes Z' and Z'' such that $Z' := (Z_1, \ldots, Z_{k-1}, [\underline{Z}_k, \text{mid } Z_k], Z_{k+1}, \ldots, Z_n)$, $Z'' := (Z_1, \ldots, Z_{k-1}, [\text{mid } Z_k, \overline{Z}_k], Z_{k+1}, \ldots, Z_n)$;

    if the equation $\phi(Z', t) = 0$ is solvable or possibly solvable on X_μ, then assign $z' := \underline{Z}'_n$ and put the record (Z', z') into $\mathcal{L}$ so that the second fields of the records in $\mathcal{L}$ increase ;

    if the equation $\phi(Z', t) = 0$ is solvable on X_μ, set $\overline{y} := \min\{ \overline{y}, \overline{Z}'_n \}$;

    if the equation $\phi(Z'', t) = 0$ is solvable or possibly solvable on X_μ, then assign $z'' := \underline{Z}''_n$ and put the record (Z'', z'') into $\mathcal{L}$ so that the second fields of the records in $\mathcal{L}$ increase ;

    if the equation $\phi(Z'', t) = 0$ is solvable on X_μ, set $\overline{y} := \min\{ \overline{y}, \overline{Z}''_n \}$;

    delete the former leading record (Z, z) from the list $\mathcal{L}$;

    denote the new leading record of the list $\mathcal{L}$ by (Z, z) ;

END DO

$\underline{y} := z$;

space $\mathbb{R}^{n+1}$, a line be given, parallel to the μ-th coordinate axis and having the parametric equation

$$
\begin{cases}
x_1 = r_1, \\
\quad \vdots \\
x_{\mu-1} = r_{\mu-1}, \\
x_\mu = t, \\
x_{\mu+1} = r_{\mu+1}, \\
\quad \vdots \\
x_n = r_n, \\
y = l,
\end{cases}
\tag{3}
$$

where t is a parameter varying over the entire of $\mathbb{R}$ and $r_1, \ldots, r_{\mu-1}, r_{\mu+1}, \ldots, r_n, l$ are some constants. Similar to the one-dimensional case,

$$
\min_{x \in X} f(x) = \min \left\{ l \in \mathbb{R} \ \middle| \ \begin{array}{l} \text{the line (3) defined by (3) intersects} \\ \text{the graph of the function } y = f(x) \end{array} \right\}
$$

provided that f is continuous. Therefore, we can "feel about" the graph of the function to be minimized by the one-dimensional lines, making use of the efficient one-dimensional interval procedures (the famous interval Newton method and modifications) to check whether the elementary "level equations" $f(x) - l = 0$ are solvable or not.

Turning to the interval optimization procedure, we designate

$$
\mathbf{Z} = (\mathbf{Z}_1, \ldots, \mathbf{Z}_n) := (\mathbf{X}_1, \ldots, \mathbf{X}_{\mu-1}, \mathbf{X}_{\mu+1}, \ldots, \mathbf{X}_n, \mathbf{Y}),
\tag{4}
$$

$$
\phi(\mathbf{Z}, t) := f(\mathbf{X}_1, \ldots, \mathbf{X}_{\mu-1}, t, \mathbf{X}_{\mu+1}, \ldots, \mathbf{X}_n) - \mathbf{Y}.
\tag{5}
$$

The n-dimensional boxes $\mathbf{Z}$ represent the bundles of straight lines parallel to the μ-th coordinate direction and "groping" the graph of the function $y = f(x)$, while the result of either intersection or nonintersection of the bundle with the graph will be determined from the solution of the one-dimensional equation $\phi(\mathbf{Z}, t) = 0$ on $\mathbf{X}_\mu$ with respect to the unknown t. Keeping all the boxes that have nonempty intersection with the graph is the guarantee that the sought-for global minimum will not be lost.

To sum up, we organize the overall process of the successive improvement of the estimates for the minimum (1) similar to what has been done in the popular "branch-and-bound" based interval global optimization techniques from [1, 3, 6]:

- we arrange all the boxes, produced from the subdivision of the initial box $\mathbf{Z}$, as a *working list* $\mathcal{L}$;
- at each step of the algorithm, the bisected box is that from the list $\mathcal{L}$ having the smallest left endpoint of the last component, i.e. the one showing the smallest estimate of the range of f;
- we bisect only the longest component in the box to be subdivided.

Additionally, the boxes of the form (4) that the list $\mathcal{L}$ consists of will be ordered so that the values of the left endpoint of their last component (they represent the ranges of values) increase. The first record of the working list is, as usual, called *leading* for the current step of the algorithm. The overall pseudocode of the new method that we are going to call *graph subdivision method* is given in Table 1, where wid means the width of an interval.

Coming up next is a more general situation when s $(1 \leq s \leq n)$ coordinate directions are declared as mute, and without loss in generality we can take the numbers of these directions as 1, 2, ..., s. Let, in the space $\mathbb{R}^{n+1}$, a plane be defined, parallel to the mute coordinate directions and thus determined by the equation

$$\begin{cases} x_1 = t_1, \\ \quad \vdots \\ x_{s-1} = t_{s-1}, \\ x_s = t_s, \\ x_{s+1} = r_{s+1}, \\ \quad \vdots \\ x_n = r_n, \\ y = l, \end{cases} \tag{6}$$

where t_1, ..., t_s are parameters varying over the whole of the real axis and r_{s+1}, ..., r_n, l are some constants. Similar to the one-dimensional case, if f is continuous on $\boldsymbol{X}$,

$$\min_{x \in \boldsymbol{X}} f(x) \;=\; \min \left\{ l \in \mathbb{R} \;\middle|\; \begin{array}{l} \text{the plane defined by (6) intersects} \\ \text{the graph of the function } y = f(x) \end{array} \right\}.$$

We denote

$$\boldsymbol{Z} = (\boldsymbol{Z}_1, \ldots, \boldsymbol{Z}_{n-s+1}) \; := \; (\boldsymbol{X}_{s+1}, \ldots, \boldsymbol{X}_n, \boldsymbol{Y}), \tag{7}$$

$$\varphi(\boldsymbol{Z}, t) \; := \; f(t_1, \ldots, t_s, \boldsymbol{X}_{s+1}, \ldots, \boldsymbol{X}_n) - \boldsymbol{Y}, \tag{8}$$

so as the $(n-s)$-dimensional boxes $\boldsymbol{Z}$ are bundles of planes of the form (6), while either the intersection or nonintersection of such bundles with the graph will be determined from the result of the solution of the interval equation $\phi(\boldsymbol{Z}, t) = 0$ with respect to $t = (t_1, t_2, \ldots, t_s)$. Therefore, we can "grope" the graph of the objective function by the planes (6) provided that we are able to effectively check the solvability of these equations of s unknowns.

Finally, we arrange the overall process of the successive improvement of the estimates for the global minimum according to the "branch-and-bound" strategy, and the pseudocode of the resulting new algorithm presented in Table 2 is quite similar to the previous case of only one mute direction.

The two above pseudocodes are evidently intended for the computation of the function's minimum (1) only, but a straightforward modification may adjust the algorithm in order to also find the values of the variables where f takes its

Table 2. The simplest graph subdivision method for global optimization * (several mute variables)

Input

A box $X \subseteq \mathbb{R}^n$ and a function $f : X \to \mathbb{R}$.   An accuracy $\epsilon > 0$.

A method for checking the solvability of the interval equation $\varphi(Z, t) = 0$
   for $t = (t_1, \ldots, t_s)$ and φ, Z, defined in (7)–(8).

Output

The lower $\underline{y}$ and upper $\overline{y}$ estimates, with the accuracy ϵ, for the global
   minimum of the function f over the box X .

Algorithm

compute an enclosure Y of the range of f over X ;

assign $Z := (X_{s+1}, \ldots, X_n, Y)$;

set $z := \underline{Y}$ and $\overline{y} := \overline{Y}$;

initialize the working list $\mathcal{L} := \{(Z, z)\}$;

DO WHILE ($\overline{y} - z \geq \epsilon$)

   choose the component k of the box Z having the largest
      length, i. e. such that wid $Z_k = \max_{1 \leq i \leq (n-s+1)}$ wid Z_i ;

   bisect the box Z along the k-th coordinate to half-boxes Z' and Z''
      such that $Z' := (Z_1, \ldots, Z_{k-1}, [\underline{Z}_k, \text{mid } Z_k], Z_{k+1}, \ldots, Z_{n-s+1})$,
         $Z'' := (Z_1, \ldots, Z_{k-1}, [\text{mid } Z_k, \overline{Z}_k], Z_{k+1}, \ldots, Z_{n-s+1})$;

   if the equation $\phi(Z', t) = 0$ on $(X_1, X_2, \ldots, X_s)$ is solvable or possibly
      solvable, then assign $z' := \underline{Z}'_{n-s+1}$ and put the record (Z', z') into $\mathcal{L}$
      so that the second fields of the records in $\mathcal{L}$ increase ;

   if the equation $\phi(Z', t) = 0$ on $(X_1, X_2, \ldots, X_s)$ is solvable, then
      set $\overline{y} := \min\{\overline{y}, \overline{Z}'_{n-s+1}\}$;

   if the equation $\phi(Z'', t) = 0$ on $(X_1, X_2, \ldots, X_s)$ is solvable or possibly
      solvable, then assign $z'' := \underline{Z}''_{n-s+1}$ and put the record (Z'', z'') into $\mathcal{L}$
      so that the second fields of the records in $\mathcal{L}$ increase ;

   if the equation $\phi(Z'', t) = 0$ on $(X_1, X_2, \ldots, X_s)$ is solvable, then
      set $\overline{y} := \min\{\overline{y}, \overline{Z}''_{n-s+1}\}$;

   delete the former leading record (Z, z) from the list $\mathcal{L}$;

   denote the new leading record of the list $\mathcal{L}$ by (Z, z) ;

END DO

$\underline{y} := z$;

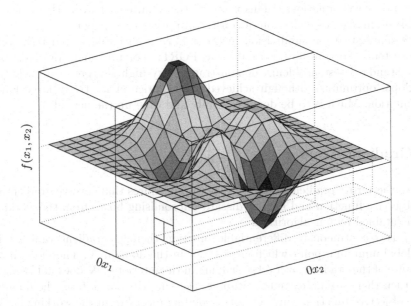

Fig. 2. A global minimization process via graph subdivision technique for an objective function $f : \mathbb{R}^2 \to \mathbb{R}$

global minimums. Namely, we should trace and store all the roots (either certain or possible) of the "level equations" $\phi(\mathbf{Z}, t) = 0$ apart from the information on their solvability. This will require extending the records that compose the working list $\mathcal{L}$ to incorporate the root enclosures into them.

What can be said about the convergence of the graph subdivision methods? In the traditional interval global optimization algorithms from [1, 3, 6], the diameters of the leading boxes are well-known to tend to zero, and this should be also valid for the graph subdivision methods inasmuch as their logical structure coincides with that of the traditional methods. Therefore, the "level equations" $\phi(\mathbf{Z}, t) = 0$ defined by (5) and (8) tend to point (noninterval) equations. If the objective function f is such that the roots of $\phi(Z, t) = 0$ depend continuously on the parameter Z, then we can expect that the graph subdivision method converges to global optimums.

Although the graph subdivision methods may appear unnecessarily complex in comparison with the traditional ("direct") interval global optimization methods based on adaptive subdivision of the domain of the objective function, there exists a large realm of problems where both approaches have equal practicalities. These are optimization problems with implicitly defined objective functions. In such problems, evaluation of the objective fucntion requires solving an equation or a system of equations anyway.

Yet another idea that can make the graph subdivision methods much more attractive and practical is the use of constraint propagation techniques for the solution of "level equations". This is one of the main reasons why the article is published among this collection of constraint propagation papers.

The simplest graph subdivision methods have been implemented using Sun Microsystems' Fortran 95 (also known as FORTE Fortran) and, for a number of the standard test problems, demonstrated very high sharpness of enclosing the global optimums, although achieved at the price of relatively large labor consumption. Much is to be done to modify and tune up the new idea.

4   Gradient Tests

We use the term "gradient tests" to denote procedures that involve gradient of the objective function and help to discard unpromising boxes from the working list maintained by our algorithm.

If f is a continuously differentiable function then its gradient vanishes in the global minimum point which are interior in the domain X. Therefore, if an enclosure of the gradient over a box x lying in the interior of X does not contain zero, then there are no extrema within x. Deleting the box x from the domain of the objective function (and the corresponding record from the working list of the algorithm) will not affect the results of the global optimization process.

If the subbox x is not interior for X, then we cannot discard it so simlply. Although the interior of x really cannot have extremums of f, one need to additionally investigate the part of x that shows up the boundary of the entire domain box X. The techniques using gradients enclosures are very popular in the interval global optimization methods (see [1, 3, 6]), but application of the above idea in the graph subdivision methods has specific character.

In the graph subdivision methods, we do not subdivide the domain along the mute coordinate directions. As the result, all the boxes from the working list intersect the boundary of the initial domain X and never become interior subboxes. We have to take this fact into account when processing new sub-boxes during the execution of the algorithm. Let, for example, the algorithm of Table 1, with the mute direction μ, has generated a record $(\boldsymbol{Z}, z)$, $\boldsymbol{Z} = (\boldsymbol{x}_1, \dots, \boldsymbol{x}_{\mu-1}, \boldsymbol{x}_{\mu+1}, \dots, \boldsymbol{x}_n, \boldsymbol{y})$, such that within the box $(\boldsymbol{x}_1, \dots, \boldsymbol{x}_{\mu-1}, X_\mu, \boldsymbol{x}_{\mu+1}, \dots, \boldsymbol{x}_n) \subseteq \boldsymbol{X}$ the gradient of the objective function does not contain zero. Hence, the sought-for extremum can be attained only at the points from the box $(\boldsymbol{x}_1, \dots, \boldsymbol{x}_{\mu-1}, X_\mu, \boldsymbol{x}_{\mu+1}, \dots, \boldsymbol{x}_n)$ that goes out to the boundary ∂X of the initial box X, i.e. they are in the intersection

$$\boldsymbol{X} \cap (\boldsymbol{x}_1, \dots, \boldsymbol{x}_{\mu-1}, X_\mu, \boldsymbol{x}_{\mu+1}, \dots, \boldsymbol{x}_n). \tag{9}$$

Therefore, at best we have to retain for further processing only two $(n-1)$-dimensional subboxes of $\boldsymbol{X}$, i.e.

$$(\boldsymbol{x}_1, \dots, \boldsymbol{x}_{\mu-1}, \underline{X}_\mu, \boldsymbol{x}_{\mu+1}, \dots, \boldsymbol{x}_n),$$
$$(\boldsymbol{x}_1, \dots, \boldsymbol{x}_{\mu-1}, \overline{X}_\mu, \boldsymbol{x}_{\mu+1}, \dots, \boldsymbol{x}_n),$$

obtained from $(\boldsymbol{x}_1, \ldots, \boldsymbol{x}_{\mu-1}, \boldsymbol{X}_\mu, \boldsymbol{x}_{\mu+1}, \ldots, \boldsymbol{x}_n)$ by throwing away the points of the interior of $\boldsymbol{X}$, and at worst we have to retain $(2n - 1)$ or even $2n$ faces of the intersection box (9) (in case $(\boldsymbol{x}_1, \ldots, \boldsymbol{x}_{\mu-1}, \boldsymbol{X}_\mu, \boldsymbol{x}_{\mu+1}, \ldots, \boldsymbol{x}_n) = \boldsymbol{X}$).

The same happens to the graph subdivision methods with several mute variables, with the sole difference that the number of pieces of the boundary that we have to retain may only increase.

To sum up, in the graph subdivision methods, we never discard the subboxes entirely, but always retain parts of their boundaries. In the traditional "direct" interval global optimization methods, whole subboxes may be in the interior of the initial domain $\boldsymbol{X}$, and we entirely discard them.

5   Application to Interval Band Linear Systems

The most interesting implementations of the idea of graph subdivision methods are those where one can take several mute variables and thus substantially descrease the dimension of the argument of the original optimization problem. As a practical example of such an application, we consider the problem of outer component-wise estimation of the solution set to interval linear systems with band matrices.

Solution set to the interval linear system

$$\boldsymbol{Ax} = \boldsymbol{b} \tag{10}$$

with an interval $m \times n$-matrix $\boldsymbol{A}$ and interval m-vector $\boldsymbol{b}$ is the set

$$\varXi(\boldsymbol{A}, \boldsymbol{b}) = \{\, x \in \mathbb{R}^n \mid (\exists A \in \boldsymbol{A})(\exists b \in \boldsymbol{b})(Ax = b)\,\},$$

formed by all solutions to the point linear systems $Ax = b$ for $A \in \boldsymbol{A}$ and $b \in \boldsymbol{b}$. Structure of the solution set is quite complex, and usually we confine ourselves to the problems of approximate description (estimation) of the solution set according to this or that sense. For simplicity, we consider only square systems of equations with an $n \times n$-matrix $\boldsymbol{A}$.

An important problem arising in connection with the interval linear systems (10) is that of computing outer component-wise estimates of the solution set:

> For an interval system of linear algebraic equations $\boldsymbol{Ax} = \boldsymbol{b}$
> find the estimates for $\min\{\, x_\nu \mid x \in \varXi(\boldsymbol{A}, \boldsymbol{b})\,\}$ from below $\qquad$ (11)
> and for $\max\{\, x_\nu \mid x \in \varXi(\boldsymbol{A}, \boldsymbol{b})\,\}$ from above, $\nu = 1, 2, \ldots, n$.

When speaking of the "solution of interval linear systems of equations" one often means the problem (11). In our work, we fix the index ν and concentrate on computing only $\min\{\, x_\nu \mid x \in \varXi(\boldsymbol{A}, \boldsymbol{b})\,\}$, since

$$\max\{\, x_\nu \mid x \in \varXi(\boldsymbol{A}, \boldsymbol{b})\,\} = -\min\{\, x_\nu \mid x \in \varXi(\boldsymbol{A}, -\boldsymbol{b})\,\}.$$

The matrix $\boldsymbol{A} = (\,a_{ij})$ (either point or interval) is called *band*, if there exists nonnegative integers p and q, such that $a_{ij} = 0$ for $j > i+p$ and $i > j+q$. Then

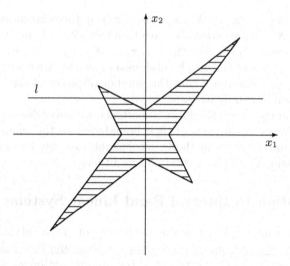

Fig. 3. Solution set and crossing it by a line

the value of $(p+q+1)$ is the *width* of the band in the matrix $\boldsymbol{A}$. Below, we take the assumption that the band is not "too wide", namely

$$p+q \leq \frac{n}{2}. \tag{12}$$

When solving the problem (11), we suppose that an interval box $\boldsymbol{V}$ is known that contains the solution set estimated, that is, $\boldsymbol{V} \supseteq \varXi(\boldsymbol{A}, \boldsymbol{b})$. The box $\boldsymbol{V}$ can be found by any of the methods described e. g. in [3–5], and its size is not a big part of the entire technique.

The fact of fundamental importance is that the problem (11) of outer interval estimation of the solution set is, in essense, an optimization problem. The corresponding reformulation can be given, for example, in the following way [8]. If $\nu \in \{1, 2, \ldots, n\}$ is a fixed index, then, through l, we denote a straight line in $\mathbb{R}^n$ that is parallel to the ν-th coordinate axis and has the parametric equation

$$\begin{cases} x_1 &=& r_1, \\ &\vdots& \\ x_{\nu-1} &=& r_{\nu-1}, \\ x_\nu &=& t, \\ x_{\nu+1} &=& r_{\nu+1}, \\ &\vdots& \\ x_n &=& r_n, \end{cases} \tag{13}$$

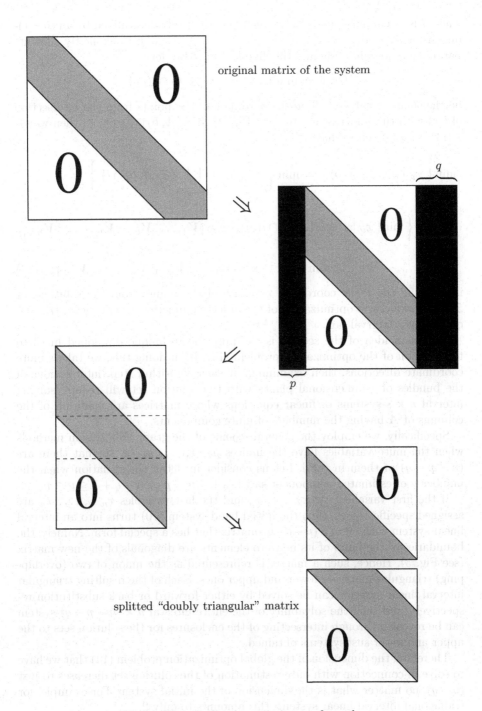

Fig. 4. How the band matrix is transformed

where t is a real parameter. Every such line is entirely determined by an $(n-1)$-dimensional vector $r = (r_1,\ldots,r_{\nu-1},r_{\nu+1},\ldots,r_n)^\top$, and, to explicitly show its parameters, we will designate this line as $l(r)$. Also, let

$$\Omega(r) \;=\; \min\{\, x_\nu \mid x \in \Xi(\boldsymbol{A},\boldsymbol{b}) \cap l(r)\,\}$$

be the smallest value of the ν-th coordinate of the points from the intersection of $l(r)$ with the solution set (10) (see Fig. 3). If $\Xi(\boldsymbol{A},\boldsymbol{b}) \cap l(r) = \varnothing$, then we set $\Omega(r) = +\infty$. Then we have

$$\min\{\, x_\nu \mid x \in \Xi(\boldsymbol{A},\boldsymbol{b})\,\} \;=\; \min\left\{\, x_\nu \;\middle|\; x \in \bigcup_{l \cap V \neq \varnothing} (\Xi(\boldsymbol{A},\boldsymbol{b}) \cap l)\, \right\}$$

$$=\; \min\left\{\, \min\{\, x_\nu \mid x \in \Xi(\boldsymbol{A},\boldsymbol{b}) \cap l(r)\,\} \;\middle|\; r \in (\boldsymbol{V}_1,\ldots,\boldsymbol{V}_{\nu-1},\boldsymbol{V}_{\nu+1},\ldots,\boldsymbol{V}_n)\, \right\}$$

$$=\; \min\{\, \Omega(r) \mid r \in (\boldsymbol{V}_1,\ldots,\boldsymbol{V}_{\nu-1},\boldsymbol{V}_{\nu+1},\ldots,\boldsymbol{V}_n)\,\}, \quad (14)$$

i.e. finding the ν-the coordinate estimate of the points from the solution set $\Xi(\boldsymbol{A},\boldsymbol{b})$ reduces to optimization of the objective function $\Omega(r)$ over an $(n-1)$-dimensional interval box.

The main idea of the section is to apply the technique developed in §3 to the solution of the optimization problem (14). If, in doing this, we take s mute coordinate directions, then examining, at each algorithm step, intersections of the bundles of s-dimensional planes with the solution set will require solving interval $n \times s$-systems of linear equations whose matrices are made up of the columns of $\boldsymbol{A}$ having the numbers of mute components.

Specifically, we employ the general scheme of the graph subdivision methods when the mute variables have the indices $p+1$, ..., $n-q$, so that there are $(n-p-q)$ of them in total. Let us consider in detail the situation when the number ν of estimated component satisfies $1 \le \nu \le p$ or $n-q+1 \le \nu \le n$.

If the first variables x_1, x_2, ..., x_p and the last variables x_{n-q}, ..., x_n are assigned specific values, then the initial band system (10) turns into an interval linear system with an $m \times (n-p-q)$-matrix that has a special form. Namely, the boundaries of the band of its nonzero elements are diagonals of the new matrix (see Fig. 3). Hence, such a matrix is represented as the union of two (overlapping) triangular matrices, lower and upper ones. Each of the resulting triangular interval linear systems can be solved by either forward or back substitution respectively, and then the solvability of the entire interval $n \times (n-p-q)$-system can be revealed through intersecting of the enclosures for the solution sets to the upper and lower susbsystems obtained.

Therefore, the dimension of the global optimization problem (14) that we have to solve in connection with outer estimation of the solution set decreases to just $(p+q)$, no matter what is the dimension of the initial system. For example, for tridiagonal interval linear systems this amounts to only 2.

In Tables 3–4, the overall algorithm for solving interval linear band systems is presented. Table 3 shows how checking solvability of the interval subsystems

Table 3. Checking solvability of subsystem generated by the algorithm of Table 4

```
DO  i = 1  TO  n
```
$$\breve{b}_i(Z) := b_i - \sum_{j=1}^{p} a_{ij} Z_j - \sum_{j=p+1}^{p+q} a_{i,j+n-p-q} Z_j$$
```
END DO
```
$$G_{p+1} := \breve{b}_1(Z) / a_{1,p+1};$$
```
DO  i = p + 2  TO  n - q
```
$$G_i := \left(\tilde{b}_{i-p}(Z) - \sum_{j=p+1}^{i-1} a_{i-p,j} G_j \right) \Big/ a_{i-p,i}$$
```
END DO
```
$$H_{n-q} := \breve{b}_n(Z) / a_{n,n-q};$$
```
DO  i = n - q - 1  DOWNTO  p + 1
```
$$H_i := \left(\tilde{b}_{i+q}(Z) - \sum_{j=i+1}^{n-q} a_{i+q,j} H_j \right) \Big/ a_{i+q,i}$$
```
END DO
IF ( G ∩ H ≠ ∅ ) THEN
        the system Ăx = b̆(Z) is solvable
ELSE
        the system Ăx = b̆(Z) is not solvable
END IF
```

can be organized, while Table 4 gives the general algorithm. In Tables 3–4, we denote $Z = (Z_1, Z_2, \ldots, Z_{p+q})^\top \in \mathbb{IR}^{p+q}$, and

$$\breve{A} = (a_{ij})_{j=p+1}^{n-q}, \qquad \breve{b}(Z) = (\breve{b}_1(Z), \breve{b}_2(Z), \ldots, \breve{b}_n(Z))^\top,$$

$$\breve{b}_i(Z) = b_i - \sum_{j=1}^{p} a_{ij} Z_j - \sum_{j=p+1}^{p+q} a_{i,j+n-p-q} Z_j.$$

The interval linear system $\breve{A}x = \breve{b}(Z)$ is an analogue of the "level equation" from §§2–3. To examine its solvability, we split it as shown in Fig. 4, and then compute the interval vectors $G = (G_1, G_2, \ldots, G_n)^\top$ and $H = (H_1, H_2, \ldots, H_n)^\top$, interval hulls of the solution sets to the lower and upper triangular interval linear systems obtained from $\breve{A}x = \breve{b}(Z)$. In the pseudocode of Table 3, G and H are found by forward substitution and back substitution respectively.

Of course, in such testing solvability of the interval system, we allow some coarsening, since we intersect not the solution sets of the subsystems, but their

Table 4. Estimating the solution set for band interval linear systems

---

Input

An interval linear system $\boldsymbol{A}x = \boldsymbol{b}$ with a band matrix $\boldsymbol{A}$.

A number ν of the estimated component of the solution set.

An interval enclosure $\boldsymbol{V} \supseteq \Xi(\boldsymbol{A}, \boldsymbol{b})$ for the solution set estimated.

Output

An estimate $\underline{y}$ for $\min\{\, x_\nu \mid x \in \Xi(\boldsymbol{A}, \boldsymbol{b}) \,\}$ from below.

Algorithm

assign $\boldsymbol{Z} := (\boldsymbol{V}_1, \ldots, \boldsymbol{V}_p, \boldsymbol{V}_{n-q+1}, \ldots, \boldsymbol{V}_n)$;

set $z := \underline{\boldsymbol{V}}_\nu$;

initialize the working list $\mathcal{L} := \{\, (\boldsymbol{Z}, z) \,\}$;

DO WHILE (the box $\boldsymbol{Z}$ is wide)

    choose the component k along which the box $\boldsymbol{Z}$ has the largest width, i. e. wid $\boldsymbol{Z}_k = \max_{1 \le i \le (p+q)}$ wid $\boldsymbol{Z}_i$;

    bisect the box $\boldsymbol{Z}$ along its k-th coordinate direction to such boxes $\boldsymbol{Z}'$ and $\boldsymbol{Z}''$ that
$$\boldsymbol{Z}' := (\boldsymbol{Z}_1, \ldots, \boldsymbol{Z}_{k-1}, [\underline{\boldsymbol{Z}}_k, \operatorname{mid} \boldsymbol{Z}_k], \boldsymbol{Z}_{k+1}, \ldots, \boldsymbol{Z}_{p+q}),$$
$$\boldsymbol{Z}'' := (\boldsymbol{Z}_1, \ldots, \boldsymbol{Z}_{k-1}, [\operatorname{mid} \boldsymbol{Z}_k, \overline{\boldsymbol{Z}}_k], \boldsymbol{Z}_{k+1}, \ldots, \boldsymbol{Z}_{p+q});$$

    if the system $\check{\boldsymbol{A}}x = \check{\boldsymbol{b}}(\boldsymbol{Z}')$ is solvable, then assign $z' := \underline{\boldsymbol{Z}}'_\nu$ and put the pair $(\boldsymbol{Z}', z')$ into $\mathcal{L}$ so that the second field of the pairs in $\mathcal{L}$ increase;

    if the system $\check{\boldsymbol{A}}x = \check{\boldsymbol{b}}(\boldsymbol{Z}'')$ is solvable, then assign $z'' := \underline{\boldsymbol{Z}}''_\nu$ and put the pair $(\boldsymbol{Z}'', z'')$ into $\mathcal{L}$ so that the second field of the pairs in $\mathcal{L}$ increase;

    delete the former leading box $(\boldsymbol{Z}, z)$ from the list $\mathcal{L}$;

    denote the current leading box of the list $\mathcal{L}$ through $(\boldsymbol{Z}, z)$;

END DO

$\underline{y} := z$;

---

interval hulls (i. e., the tightest enclosures), that is, wider sets. Let us consider, as an example, the interval linear system

$$\begin{pmatrix} 1 & 1 \\ 0 & 1 \end{pmatrix} \cdot \begin{pmatrix} x_1 \\ x_2 \end{pmatrix} = \begin{pmatrix} [-1, 1] \\ [-1, 1] \end{pmatrix}. \tag{15}$$

Its matrix is an upper triangular point matrix, and the solution set is depicted at Fig. 5. As the result, we can compute an estimate of the solution set to the band interval linear system which is not optimal. But our computational experience

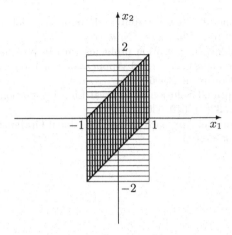

Fig. 5. Solution set for the system (15) and its interval hull

shows that they are quite sharp providing that the band width is small and intervals in the matrix are not too wide.

As distinct from the graph subdivision methods from §§3–4 designed for the solution of general optimization problems, we do not have to involve interval bounds on the mute variables, that is, V_{p+1}, V_{p+2}, ..., V_{n-q}. The point is that the procedure for testing solvability of the "level equations" used in the algorithm (Table 3) can spare these values. Hence, initially we suffice to know not the entire box $V \supseteq \Xi(A, b)$, but only its components V_1, ..., V_p and V_{n-q+1}, ..., V_n.

There is room for further improvement of our algorithm through taking into account fine geometric structure of the solution set to triangular interval linear systems. For example, we can use prisms for enclosing them rather than axis aligned boxes. This will require additional efforts to reveal their intersection, but results in sharper estimates of the solution set.

References

1. Hansen, E., Walster, G.W.: Global Optimization Using Interval Analysis. Marcel Dekker, New York (2004)
2. Kearfott, R.B., Nakao, M.T., Neumaier, A., Rump, S.M., Shary, S.P., van Hentenryck, P.: Standardized Notation in Interval Analysis. Computational Technologies 15(1), 7–13 (2010)
3. Kearfott, R.B.: Rigorous Global Search. Continuous Problems. Kluwer, Dordrecht (1996)
4. Moore, R.E., Kearfott, R.B., Cloud, M.J.: Introduction to Interval Analysis. SIAM, Philadelphia (2009)
5. Neumaier, A.: Interval Methods for Systems of Equations. Cambridge University Press, Cambridge (1990)

6. Ratschek, H., Rokne, J.: New Computer Methods for Global Optimization. Ellis Horwood, Halsted Press, Chichester, New York (1988)
7. Semenov, A.L.: Solving optimization problems with help of the UniCalc solver. In: Kearfott, R.B., Kreinovich, V. (eds.) Applications of Interval Computations, pp. 211–225. Kluwer, Dordrecht (1996)
8. Shary, S.P.: On Optimal Solution of Interval Linear Equations. SIAM Journal on Numerical Analysis 32(2), 610–630 (1995)
9. Shary, S.P.: A Surprising Approach in Interval Global Optimization. Reliable Computing 7(6), 497–505 (2001)

An Extended BDI-Based Model for Human Decision-Making and Social Behavior: Various Applications

Young-Jun Son

Systems and Industrial Engineering
The University of Arizona
Tucson, AZ 85721, USA
son@sie.arizona.edu

Abstract. An extended Belief-Desire-Intention (BDI) modeling framework has been developed and refined by the author's research group in the last decade to mimic realistic human decision-making and social behaviors. The goal of this manuscript is to discuss various applications that the proposed modeling framework has been applied, such as 1) evacuation behaviors under a terrorist bomb attack, 2) pedestrian behaviors in the Chicago Loop area, 3) workforce assignment in a multi-organizational social network for community-based software development, 4) pedestrian behaviors in a shopping mall, 5) evacuation behaviors under fire in a factory, and 6) error detection and resolution by people in a complex manufacturing facility.

Keywords: BDI, human decision behavior, planning, Bayesian belief network.

1 Introduction

The goal of this manuscript is to describe an extended Belief-Desire-Intention (BDI) modeling framework [4] and [5] (see Fig. 1(a)) that has been developed and refined by the author's research group in the last decade for human decision-making and social behavior, effectively integrating engineering-, psychology-, and economics-based models. This manuscript is a reproduced version of the extended abstract that was contained in the Proceedings of CoProd'11.

BDI [1, 6] is a model of the human reasoning process, where a person's mental state is characterized by three major components: beliefs, desires, and intentions. Later, Zhao and Son [10] extended the decision-making module (corresponding to the intention component) of the original BDI model to include three detailed submodules: (1) a deliberator, (2) a real-time planner, and (3) a decision executor in the decision-making (intention) module, where this extension was necessary to accommodate both the decision-making and decision-planning functions in a unified framework. In addition, an emotional module containing a confidence index and time pressure also has been appended to represent these aspects of

M. Ceberio and V. Kreinovich (eds.), *Constraint Programming and Decision Making*,
Studies in Computational Intelligence 539,
DOI: 10.1007/978-3-319-04280-0_19, © Springer International Publishing Switzerland 2014

human psychology. The emotional module affects and is affected by the three other mental modules, that is, beliefs, desires, and decision making. While Zhao and Son [10] provided a conceptual extension of the BDI model, Lee and Son [4] later proposed actual algorithms and techniques that have been employed and further developed to realize submodules for the extended model.

The submodules of the extended BDI modeling framework proposed by Lee and Son [4, 5] are based on a Bayesian Belief Network (BBN), Decision-Field-Theory (DFT), and a Probabilistic Depth-First Search (PDFS) technique, and a key novelty of the framework is its ability to represent both the human decision-making and decision-planning functions in a unified framework.

The extended BDI modeling framework mentioned in the previous in the previous paragraphs has been sucessfully demonstrated for a human's behaviors under various applications, such as 1) evacuation behaviors under a terrorist bomb attack (see Fig. 1(b)), 2) workforce assignment in a multi-organizational social network for community-based software development (see Fig. 1(c)), 3) pedestrian behaviors in the Chicago Loop area (see Fig. 1(d)), 4) pedestrian behaviors in a shopping mall (see Fig. 1(e)), 5) evacuation behaviors under fire in a factory (see Fig. 1(f)), 6) error detection and resolution by people in a complex manufacturing facility, and 7) driver's behaviors in path/route selection [3].

To mimic realistic human behaviors, attributes of the extended BDI framework are reverse-engineered from human-in-the-loop experiments conducted in the Cave Automatic Virtual Environment (CAVE) or other simulated settings (e.g. software simulator running in a desktop or driving simulators). For emergency evacuation scenario as an example [4, 5], each subject who participated in the experiment was asked to evaluate the risk and the evacuation time of three available paths (i.e., right, forward, and left) depending on the various environmental observations (i.e., fire, smoke, police, and crowd) at each intersection. Also, each subject was asked to select one of the three available paths. The data collected on the relationship between the environment and the subject's evaluation was used to construct a BBN in the form of a conditional probability distribution. The constructed BBN infers 1) subjective evaluations for each attribute (e.g., risk and time) of each given option and 2) subjective weights of attention corresponding to each attribute, and the DFT calculates preference values of the options based on those matrices of evaluations and weights.

After each agent model conforming to the extended BDI-framework is calibrated with the data collected from the human-in-the loop experiment, agents together with the simulated environment (e.g. buildings, fire, smoke) are implemented. In our work, AnyLogic (http://www.anylogic.com/) and Repast (http://repast.sourceforge.net/) have been mainly used for those various applications depicted in Fig. 1. For each of the considered applications, the constructed simulation has been then used to test the impact of various control factors (e.g. demographics, number of police officers, information sharing via speakers for the case of emergency evacuation) on the performance of interest (e.g. average evacuation time, percentage of casualties).

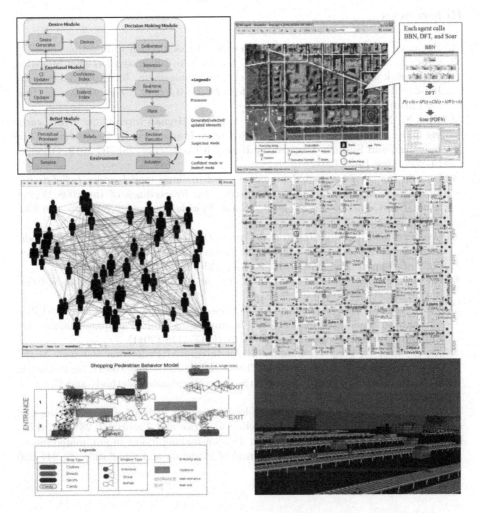

Fig. 1. (a) Components of an extended BDI framework [4, 5]; (b) Snapshot of emergency evacuation simulation [5]; (c) Snapshot of multi-organizational social network simulation [2]; (d) Snapshot of pedestrian behaviors in Chicago Loop area [9]; (e) Snapshot of pedestrian behaviors in a shopping mall [8]; (f) Snapshot of evacuation behaviors under fire in factory [7]

Later, the author's group has been further refining their extended BDI framework to address learning/forgetting of human as well as major human interactions (e.g. avoidance, accommodation, compromise, collaboration and competition). A preliminary work is available in [4], which is being researched further by the author's group.

References

1. Bratman, M.: Intention, Plans, and Practical Reason. CSLI Publications (1987)
2. Celik, N., Lee, S., Xi, H., Mazhari, E., Son, Y.: An Integrated Decision Modeling Framework for Multi-Organ. Social Network Management. In: Proceedings of the Industrial Engineering Research Conference, IERC 2010, Cancun, Mexico, June 5-9 (2010)
3. Kim, S., Xi, H., Mungle, S., Son, Y.: Modeling Human Interactions with Learning under the Extended Belief-Desire-Intention Framework. In: Proceedings of the Industrial and Systems Engineering Research Conference, ISERC 2012, Orlando, Florida, May 19-23 (2012)
4. Lee, S., Son, Y.: Integrated human decision making model under belief-desire-intention framework for crowd simulation. In: Proceedings of the 2008 Winter Simulation Conference WSC 2008, Miami, Florida, December 7-10, pp. 886–894. IEEE, New Jersey (2008)
5. Lee, S., Son, Y., Jin, J.: Integrated Human Decision Making and Planning Model under Extended Belief-Desire-Intention Framework. ACM Transactions on Modeling and Computer Simulation 20(4), 23(1)–23(24) (2010)
6. Rao, A., Georgeff, M.: Decision procedures for BDI logics. Journal of Logic Computing 8, 293–343 (1998)
7. Vasudevan, K., Son, Y.: Concurrent Consideration of Evacuation Safety and Productivity in Manufacturing Facility Planning using Multi-Paradigm Simulations. Computers and Industrial Engineering 61, 1135–1148 (2011)
8. Xi, H., Lee, S., Son, Y.: Chapter 4: An Integrated Pedestrian Behavior Model Based on Extended Decision Field Theory and Social Force Model. In: Rothrock, L., Narayanan, S. (eds.) Human-in-the-Loop Simulation: Methods and Practice. Springer (2011)
9. Xi, H., Son, Y.: Two-Level Modeling Framework for Pedestrian Route Choice and Walking Behaviors. Simulation Modelling Practice and Theory 22, 28–46 (2012)
10. Zhao, X., Son, Y.: BDI-based human decision-making model in automated manufacturing systems. International Journal of Modeling and Simulation 28, 347–356 (2008)

Why Curvature in L-Curve:
Combining Soft Constraints

Uram Anibal Sosa Aguirre, Martine Ceberio, and Vladik Kreinovich

Computational Sciences Program and Department of Computer Science
University of Texas, El Paso, TX 79968, USA
usosaaguirre@miners.utep.edu, {mceberio,vladik}@utep.edu

Abstract. In solving inverse problems, one of the successful methods of determining the appropriate value of the regularization parameter is the *L-curve method* of combining the corresponding soft constraints, when we plot the curve describing the dependence of the logarithm x of the mean square difference on the logarithm y of the mean square non-smoothness, and select a point on this curve at which the curvature is the largest. This method is empirically successful, but from the theoretical viewpoint, it is not clear why we should use curvature and not some other criterion. In this paper, we show that reasonable scale-invariance requirements lead to curvature and its generalizations.

Keywords: soft constraints, inverse problems, regularization, L-curve, curvature.

1 Formulation of the Problem

Inverse Problem: A Brief Reminder. In science and engineering, we are interested in the state of the world, i.e., in the values of different physical quantities that characterize this state. Some of these quantities we can directly measure, but many quantities are difficult or even impossible to measure directly.

For example, in geophysics, we are interested in the density and other properties of the material at different depths and different locations. In principle, it is possible to drill a borehole and directly measure these properties, but this is a very expensive procedure, and for larger depths, the drilling is not possible at all. To find the values of such difficult-to-measure quantities $q = (q_1, \ldots, q_n)$, we measure the values of the auxiliary quantities $a = (a_1, \ldots, a_m)$ that are related to q_i by a known dependence $a_i = f_i(q_1, \ldots, q_n)$, and then reconstruct the values q_j from these measurement results.

In the idealized situation when measurements are absolutely accurate, we can then reconstruct the desired values q_j from the system of m equations $a_1 = f_1(q_1, \ldots, q_n), \ldots, a_m = f_m(q_1, \ldots, q_n)$. In real life, measurements are never 100% accurate, so the measured values a_i are only approximately equal to $f_i(q_1, \ldots, q_n)$. Usually, it is assumed that the measurement errors $a_i - f_i(q_1, \ldots, q_n)$ are independent normally distributed random variables with 0 means and the same variance; see, e.g. [3]. In this case, the constraint that the

M. Ceberio and V. Kreinovich (eds.), *Constraint Programming and Decision Making,*   175
Studies in Computational Intelligence 539,
DOI: 10.1007/978-3-319-04280-0_20, © Springer International Publishing Switzerland 2014

values q_j are consistent with the observations a_i can be described as a constraint $s \leq s_0$ on the sum $s \overset{\text{def}}{=} \sum_{i=1}^{m} (a_i - f_i(q_1, \ldots, q_n))^2$. The value s_0 depends on the confidence level: the larger s_0, the more confident we are that this constraint will be satisfied. For each value x_0, the constraint $x \leq x_0$ is a *soft constraint*: there is a certain probability that this constraint will be violated.

Often, this constraint is described in a logarithmic scale, as $x \leq x_0$, where $x \overset{\text{def}}{=} \ln(s)$.

Regularization: How to Take into Account Additional Constraints. Often, there are additional constraints on q_j. Usually, the values q_j are more regular than randomly selected values. Methods for taking these additional regularity constraints into account are known as *regularization methods*; see, e.g., [4].

For example, in geophysics, the density values at nearby locations are usually close to each other. In other words, the differences $q_j - q_{j'}$ corresponding to nearby locations should be small.

This constraint can also be described in statistical terms: that there is a prior distribution on the set of all the tuples, in which all the differences $q_j - q_{j'}$ are independent and normally distributed with 0 mean and the same variance. In this case, the constraint that the values q_j are consistent with this prior distribution can be also described as a constraint $t \leq t_0$ on the sum $t \overset{\text{def}}{=} \sum_{(j,j')} (q_j - q_{j'})^2$. This constraint is also often described in a logarithmic space, as $y \leq y_0$, where $y \overset{\text{def}}{=} \ln(t)$.

We can combine the two constraints, e.g., by using the Bayesian statistics to combine the prior distribution (describing the regularity of the actual values) and the distribution corresponding to measurement uncertainty. For the resulting posterior distribution, the Maximum Likelihood method of determining the optimal values of the quantities q_j is then equivalent to minimizing the sum $s + \lambda \cdot t$, for some coefficient λ depending on the variance of the prior distribution.

There are also other more complex regularization techniques; see [4].

How to Determine a Regularization Parameter. As we have mentioned, the actual value of the regularization parameter depends on the prior distribution and is, therefore, reasonably subjective. It is therefore desirable to find the value of this parameter based on the data.

For each value of the parameter λ, we can find the corresponding solution $q_j(\lambda)$, and, based on this solution, compute the values $x(\lambda)$ and $y(\lambda)$ of the quantities x and y. These two values represent a point on a plane. Points corresponding to different values λ form a curve. In these terms, the question of which value λ to choose can be reformulated as which point on the curve should we choose?

In practice, often, this curve has a clear turning point, a point that is distinct from others – as a point at which the curve "curves" the most. In such cases, when we have an L-shaped curve, it is reasonable to select the turning point as the point corresponding to the solution. This idea often leads to a good solution; see, e.g., [1, 2].

In line with the above description, the desired point is selected as a point at which the absolute value $|C|$ of the curvature $C = \dfrac{x'' \cdot y' - y'' \cdot x'}{((x')^2 + (y')^2)^{3/2}}$ takes the largest possible value; here, as usual, x' denotes the derivative $\dfrac{dx}{d\lambda}$, and x'' denotes the second derivative of x with respect to the parameter λ.

Remaining Open Problem. Empirically, the method of selecting a point with the largest curvature works well. It is therefore desirable to come up with a theoretical justification for the use of curvature function – or at least for a class containing the curvature function.

What We Do in This Paper. We provide such a justification: specifically, we show that reasonable properties select a class of functions that include curvature.

2   Analysis of the Problem

Let us first analyze the invariance properties of curvature.

Scale-Invariance. The numerical values of each quantity depend on the selection of a measuring unit. For example, if instead of meters, we use centimeters, then all numerical values get multiplied by 100. In general, if we select a new measuring unit which is c times smaller than the previous one, then all numerical values get multiplied by c.

If we change a measuring unit for a to a new one which is c_a time smaller, then the numerical values of a_i and $a_i - f_i(q_1, \ldots, q_n)$ get multiplied by c_a. As a result, the sum $s = \displaystyle\sum_{i=1}^{n} (a_i - f_i(q_1, \ldots, q_n))^2$ gets multiplied by c_a^2, and the original value $x = \ln(s)$ changes to $x + \Delta_x$, where we denoted $\Delta_x \stackrel{\text{def}}{=} \ln(c_a^2)$.

Similarly, if we change a measuring unit for q to a new one which is c_q time smaller, then the numerical values of q_j and $q_j - q_{j'}$ get multiplied by c_q. As a result, the sum $t = \sum(q_j - q_{j'})^2$ gets multiplied by c_q^2, and the original value $y = \ln(t)$ changes to $y + \Delta_y$, where we denoted $\Delta_y \stackrel{\text{def}}{=} \ln(c_q^2)$.

Under these changes $x(\lambda) \to x(\lambda) + \Delta_x$ and $y(\lambda) \to y(\lambda) + \Delta_y$, the derivatives do not change – since Δ_x and Δ_y are constants – and thus, the curvature does not change. Thus, the curvature is invariant under these scale transformations.

Invariance under Re-scaling of Parameters. Instead of the original parameter λ, we can use a new parameter μ for which $\lambda = g(\mu)$. This re-scaling of a parameter does not change the curve itself and thus, does not change its curvature. So, the curvature is invariant under these scale transformations.

Our Idea. Our main idea is to describe all the functions which are invariant with respect to both types of re-scalings.

3   Main Result

Definition. *By* a parameter selection criterion *(or simply* criterion, *for short), we mean a function $F(x, y, x', y', x'', y'')$ of six variables. We say that the parameter selection criterion $F(x, y, x', y', x'', y'')$ is:*

  - scale-invariant *if for all possible values Δ_x and Δ_y, we have*

$$F(x + \Delta_x, y + \Delta_y, x', y', x'', y'') = F(x, y, x', y', x'', y'');$$

  - invariant w.r.t. parameter re-scaling *if for every function $g(z)$ and for the functions $\widetilde{x}(\mu) = x(g(\mu))$ and $\widetilde{y}(\mu) = y(g(\mu))$, we have*

$$F(\widetilde{x}, \widetilde{y}, \widetilde{x}', \widetilde{y}', \widetilde{x}'', \widetilde{y}'') = F(x, y, x', y', x'', y'').$$

Notation. By $C(x, y, x', y', x'', y'')$, we denote the parameter selection criterion corresponding to curvature.

Comment. Once a criterion is selected, for each problem, we use the value λ for which the value $F(x(\lambda), y(\lambda), x'(\lambda), y'(\lambda), x''(\lambda), y''(\lambda))$ is the largest.

Main Result. *A parameter selection criterion which is scale-invariant and invariant w.r.t. parameter re-scaling if and only if it has the form*

$$F(x, y, x', y', x'', y'') = f\left(C(x, y, x', y', x'', y''), \frac{x'}{y'}\right)$$

for some function $f(C, z)$.

Proof

1°. For each tuple (x, y, x', y', x'', y''), by taking $\Delta_x = -x$ and $\Delta_y = -y$, we conclude that $F(x, y, x', y', x'', y'') = F(0, 0, x', y', x'', y'')$. Thus, we conclude that $F(x, y, x', y', x'', y'') = F_0(x', y', x'', y'')$, where we denoted $F_0(x', y', x'', y'') \stackrel{\text{def}}{=} F(0, 0, x', y', x'', y'')$, i.e., we conclude that the value of the parameter selection criterion does not depend on x and y at all.

In terms of the function F_0, invariance w.r.t. parameter re-scaling means that $F_0(\widetilde{x}', \widetilde{y}', \widetilde{x}'', \widetilde{y}'') = F_0(x', y', x'', y'')$.

2°. When we go from the original function $x(\lambda)$ to the new function $\widetilde{x}(\mu) = x(g(\mu))$, the chain rule for differentiation leads to $\widetilde{x}' = x' \cdot g'$ and thus, $\widetilde{x}'' = x'' \cdot (g')^2 + x' \cdot g''$. Similarly, $\widetilde{y}' = y' \cdot g'$ and $\widetilde{y}'' = y'' \cdot (g')^2 + y' \cdot g''$.

In particular, at the point where $g' = 1$, we have $\widetilde{x}' = x$, $\widetilde{x}'' = x'' + x' \cdot g''$, $\widetilde{y}' = y'$, and $\widetilde{y}'' = y'' + y' \cdot g''$, and thus, invariance w.r.t. parameter re-scaling means that $F_0(x', y', x'' + x' \cdot g'', y'' + y' \cdot g'') = F_0(x', y', x'', y'')$. This is true for every possible values of g''. In particular, for $g'' = -\dfrac{y''}{y'}$, we have $y'' + y' \cdot g'' = 0$ and thus,

$$F_0(x', y', x'', y'') = F_0\left(x', y', x'' - x' \cdot \frac{y''}{y'}, 0\right).$$

Since

$$x'' - x' \cdot \frac{y''}{y'} = C \cdot \frac{((x')^2 + (y')^2)^{3/2}}{y'},$$

we thus conclude that

$$F_0(x', y', x'', y'') = h(C, x', y'),$$

where

$$h(C, x', y') \stackrel{\text{def}}{=} F_0\left(x', y', C \cdot \frac{((x')^2 + (y')^2)^{3/2}}{y'}, 0\right).$$

For the new function $h(C, x', y')$, since the curvature is invariant w.r.t. parameter re-scaling, invariance means that $h(C, \tilde{x}', \tilde{y}') = h(C, x', y')$. This means that

$$h(C, x', y') = h(C, x' \cdot g', y' \cdot g').$$

This is true for every possible values of g'. In particular, for $g' = \dfrac{1}{x'}$, we have $x' \cdot g' = 1$ and thus,

$$F(x, y, x', y', x'', y'') = F_0(x', y', x'', y'') = h(C, x', y') = h\left(C, 1, \frac{y'}{x'}\right),$$

i.e., $F(x, y, x', y', x'', y'') = f\left(C, \dfrac{y'}{x'}\right)$ for $f(C, z) \stackrel{\text{def}}{=} h(C, 1, z)$.

The statement is proven.

Acknowledgments. This work was supported in part by the National Science Foundation grants HRD-0734825 and DUE-0926721 and by Grant 1 T36 GM078000-01 from the National Institutes of Health.

References

1. Hansen, P.C.: Analysis of discrete ill-posed problems by means of the L-curve. SIAM Review 34(4), 561–580 (1992)
2. Moorkamp, M., Jones, A.G., Fishwick, S.: Joint inversion of receiver functions, surface wave dispersion, and magnetotelluric data. Journal of Geophysical Research 115, B04318 (2010)
3. Rabinovich, S.: Measurement Errors and Uncertainties: Theory and Practice. American Institute of Physics, New York (2005)
4. Tikhonov, A.N., Arsenin, V.Y.: Solutions of Ill-Posed Problems. W. H. Whinston & Sons, Washington, D.C. (1977)

Surrogate Models for Mixed Discrete-Continuous Variables

Laura P. Swiler[1], Patricia D. Hough[1], Peter Qian[2], Xu Xu[2], Curtis Storlie[3], and Herbert Lee[4]

[1] Sandia National Laboratories,
Albuequerque, New Mexico, and Livermore, California
{lpswile,pdhough}@sandia.gov
[2] University of Wisconsin-Madison, Madison, Wisconcin
{peterq,xuxu}@stat.wisc.edu
[3] Los Alamos National Laboratory, Los Alamos, New Mexico
storlie@lanl.gov
[4] Univerity of California, Santa Cruz, Santa Cruz, California
herbie@soe.ucsc.edu

Abstract. Large-scale computational models have become common tools for analyzing complex man-made systems. However, when coupled with optimization or uncertainty quantification methods in order to conduct extensive model exploration and analysis, the computational expense quickly becomes intractable. Furthermore, these models may have both continuous and discrete parameters. One common approach to mitigating the computational expense is the use of response surface approximations. While well developed for models with continuous parameters, they are still new and largely untested for models with both continuous and discrete parameters. In this work, we describe and investigate the performance of three types of response surfaces developed for mixed-variable models: Adaptive Component Selection and Shrinkage Operator, Treed Gaussian Process, and Gaussian Process with Special Correlation Functions. We focus our efforts on test problems with a small number of parameters of interest, a characteristic of many physics-based engineering models. We present the results of our studies and offer some insights regarding the performance of each response surface approximation method.

1   Introduction

Large-scale computational models have become common tools for analyzing complex man-made systems. We are particularly interested in engineering models that have a small number of variables of interest but are characterized by computationally expensive equation solvers such as partial differential equation solvers. Many methods for exploring such models with data and scenario uncertainties are computationally expensive. One key to successfully mitigating the computational expense involves the construction of surrogates for the large-scale models. A surrogate can take many forms, but in this context we mean a meta-model or

M. Ceberio and V. Kreinovich (eds.), *Constraint Programming and Decision Making*,
Studies in Computational Intelligence 539,
DOI: 10.1007/978-3-319-04280-0_21, © Springer International Publishing Switzerland 2014

response surface approximation built from a limited amount of data generated by the computationally expensive model. The purpose of the surrogate model is to increase the efficiency of analyses that require frequent model interrogations such as optimization and uncertainty quantification.

When the models are computationally demanding, meta-model approaches to their analysis have been shown to be very useful. For example, one standard approach in the literature is to develop an emulator that is a stationary smooth Gaussian process [13, 23, 25]. There are many good overview articles which compare various metamodel strategies. For example, Storlie et al. compare various smoothing predictors and nonparametric regression approaches in [30, 31]. Simpson et al. provide an excellent overview not just of various statistical meta-model methods but also approaches which use low-fidelity models as surrogates for high fidelity models [27]. This paper also addresses the use of surrogates in design optimization, which is a popular research area for computationally expensive disciplines such as computational fluid dynamics in aeronautical engineering design. Haftka and his students developed an approach which uses "ensembles" of emulators or hybrid emulators [35, 36]. The advantage of these types of hybrid or ensembles of emulators is that better performance can be obtained. For example, one can select the best surrogate for various features or responses, or one can use weighted model averaging of surrogates.

The particular challenge we address in our work is that of assessing the accuracy relative to computational cost of surrogates for models that have both continuous and discrete variables. While historically the variables of interest in engineering models have been continuous, there is a growing use of discrete variables that represent modeling choices (alternative plausible models) and design choices (e.g. discrete choices of materials, components, or operational settings). The major challenge in using surrogates for mixed variable problems is in handling the discrete variables. Typically, in surrogate models constructed over continuous variables (e.g. polynomial regression, splines, Gaussian process models), there is the assumption of continuity: as a continuous variable varies by a small amount, the response is assumed to vary smoothly. With discrete variables, we do not necessarily have continuous behavior. For example, if a discrete variable representing a design choice varies from choice A to choice B, the system may respond in a fundamentally different manner. Thus, surrogate modeling approaches generally do not explicitly allow for categorical input variables. One option is to order these categorical inputs in some way and treat them as continuous variables when creating a metamodel. In some cases, this can lead to undesirable and misleading results. The other option is categorical regression. In this approach, a separate surrogate model is constructed over the continuous variables for each possible combination of the discrete variable values. This approach has the advantage that the surrogate is only constructed on the continuous variables, conditional on a particular combination of discrete values. However, it quickly becomes infeasible as one increases the number of discrete variables and/or the number of "levels" per variable [18]. It is clear that a more appropriate and efficient treatment of categorical inputs is needed.

In this work, we consider three approaches in the literature for constructing mixed variable surrogates. They have their roots in response surface modeling for continuous problems and tractably incorporate discrete variables in a manner that relies on some simplifications and additional assumptions. Our goal is to empirically evaluate and compare the three methods in order to gain insight into how problem characteristics influence the efficacy of each surrogate approach.

The remaining sections of this paper discuss the three approaches we evaluated and the outcome of our computational experiments. Section 2 outlines three approaches for constructing mixed variable surrogates. Section 3 describes our approach for testing and evaluating the three surrogate methods. Section 4 provides results of the surrogates on several test problems, and Section 5 summarizes the outcome.

2   Mixed Surrogate Approaches

This section describes three classes of methods that we investigated to generate surrogate models for mixed discrete-continuous variable problems. These three classes of methods are:

ACOSSO: ACOSSO, the Adaptive COmponent Selection and Smoothing Operator, is a specialized smoothing spline model [29]. It uses the smoothing spline ANOVA decomposition to separate the underlying function into simpler functional components (i.e., main effects, two-way interactions, etc.) then explicitly estimates these functional components in one optimization. The estimation proceeds by optimizing the likelihood subject to a penalty on each of the functional components. Each component involving continuous predictors has a penalty on its roughness and overall trend, each component involving discrete predictors has a penalty on its magnitude (L^2-norm), while interaction components involving both discrete and continuous predictors receive a combination of these penalties.

Gaussian Processes with Special Correlation Functions: Gaussian process models are powerful emulators for computer models. A Gaussian process model is defined by its mean and covariance function. The covariance function specifies how the response between two points is related: the idea is that points close together in input space will tend to have responses that are similar. Typically, the covariance function is a function of the distance between the points. We investigated a variety of covariance functions that represent the covariance between discrete points and that appropriate for mixed variable problems, all developed in Qian et al. [20, 39]: the exchangeable correlation (EC), the multiplicative correlation (MC), and the unrestricted correlation function (UC). For comparison, we also looked at the Individual Kriging (IK) model which involves constructing a separate Gaussian process surrogate over the continuous variables for each combination of discrete variables. This is similar to categorical regression.

TGP: TGP, the treed Gaussian Process model, is an approach which allows different Gaussian process models (GPs) to be constructed on different partitions of the space [8, 9]. This approach naturally lends itself to discrete variables, where the partitioning can be done between different values or sets of discrete variables. In TGP, the discrete or categorical variables are converted to a series of binary variables. The binary variables are then what are partitioned upon: they become the "nodes" of the tree [10].

2.1 Adaptive COmponent Selection and Shrinkage Operator (ACOSSO)

The Adaptive COmponent Selection and Shrinkage Operator (ACOSSO) estimate [29] was developed under the smoothing spline ANOVA (SS-ANOVA) modeling framework. As it is a smoothing type method, ACOSSO works best when the underlying function is somewhat smooth. The type of splines we are using involve the minimization of an objective function involving a sum-of-squares error term, similar to regression modeling. However, in the objective function for the splines, there are additional terms which can be viewed as regularization terms: these penalty terms help smooth the function and they also help perform variable selection. In the ACOSSO implementation, there is a penalty on functions of the categorical predictors. This penalty formulation provides a variable selection and automatic model reduction: it encourages some of the terms in the objective function to be zero, removing certain discrete variables or levels of discrete variables from the formulation. To facilitate the description of ACOSSO, we first describe the multiple smoothing spline, then introduce the treatment of categorical predictors and the ACOSSO estimator.

Multivariate Smoothing Splines. Consider a vector of predictors $\mathbf{x} = [x_1, \ldots, x_I]'$. Assume that the unknown function f to be estimated belongs to 2nd order Sobolev space $\mathcal{S}^2 = \{f : f, f'$ are absolutely continuous and $f'' \in \mathcal{L}^2[0,1]\}$. The simplest extension of smoothing splines to multiple inputs is the additive model [12]. For instance, assume that

$$f \in \mathcal{F}_{add} = \{f : f(\mathbf{x}) = \sum_{i=1}^{I} g_i(x_i), g_i \in \mathcal{S}^2\}, \tag{1}$$

i.e., $f(\mathbf{x}) = \sum_{i=1}^{I} g_i(x_i)$ is a sum of univariate functions. Let $\mathbf{x}_n = [x_{n,1}, \ldots, x_{n,I}]'$ be the nth observation of a multivariate predictor $\mathbf{x}$, $n = 1, \ldots, N$, and $y_n = f(\mathbf{x}_n) + \varepsilon_n$. The additive smoothing spline estimate of f is the minimizer of

$$\frac{1}{n} \sum_{n=1}^{N} [y_n - f(\mathbf{x}_n)]^2 + \sum_{i=1}^{I} \lambda_i \int_0^1 [g_i''(x_i)]^2 \, dx_i \tag{2}$$

over $f \in \mathcal{F}_{add}$. The minimizer of the expression in Eq. (2), $\hat{f}(\mathbf{x}) = \sum_{i=1}^{I} \hat{g}_i(x_i)$, takes the form of a natural cubic spline for each of the functional components $\hat{g}_i$. Notice that there are I tuning parameters (λ_i) for the additive smoothing spline. These tuning parameters λ_i control the trade-off in the resulting estimate between smoothness and fidelity to the data; large values of λ will result in smoother functions while smaller values of λ result in rougher functions that more closely match the data. Generally, λ_i is chosen by generalized cross validation (GCV) [6] or m-fold CV [14]. A generalization to two-way and higher order interaction functions can also be achieved in a similar manner; see [29] for the full details of including interactions in the SS-ANOVA framework. The minimizer of the expression in Eq. (2) can be obtained in an efficient manner via matrix algebra using results from reproducing kernel Hilbert space (RKHS) theory; for details see [37] or [11].

Discrete Predictors. A large advantage to the SS-ANOVA framework is the ability to handle categorical predictors with relative ease. To facilitate the discussion, we generalize our notation to the following. Assume that $\mathbf{x} = [x_1, \ldots, x_I]'$ are continuous on $[0, 1]$ as previously in this section, while $\mathbf{z} = [z_1, \ldots, z_J]'$ are unordered discrete variables, and let the collection of the two types of predictors be denoted $\mathbf{w} = [\mathbf{x}', \mathbf{z}']'$. For simplicity, assume $z_j \in \{1, 2, \ldots, b_j\}$ for $j = 1, \ldots, J$ where the ordering of the integers representing the groups for z_j is completely arbitrary. For notational convenience, let $\mathcal{G}_i = \mathcal{S}^2$ for $i = 1, \ldots, I$. Also let the class of $\mathcal{L}^2$ functions on the domain of z_j (i.e., $\{1, 2, \ldots, b_j\}$) be denoted as $\mathcal{H}_j$ for $j = 1, \ldots, J$.

For simplicity, we can once again consider the class of additive functions,

$$\mathcal{F}_{add} = \{f : f(\mathbf{w}) = \sum_{i=1}^{I} g_i(x_i) + \sum_{j=1}^{J} h_j(z_j), \ g_i \in \mathcal{G}_i, \ h_j \in \mathcal{H}_j\}. \tag{3}$$

Let $\mathbf{w}_n = [x_{n,1}, \ldots, x_{n,I}, z_{n,1}, \ldots, z_{n,J}]'$ be the nth observation of a multivariate predictor $\mathbf{w}$. The traditional additive smoothing spline is then the minimizer of

$$\frac{1}{N} \sum_{n=1}^{N} [y_n - f(\mathbf{w}_n)]^2 + \sum_{i=1}^{I} \lambda_i \int_0^1 [g_i''(x_i)]^2 \, dx_i \tag{4}$$

over $f \in \mathcal{F}_{add}$. Notice that in the traditional smoothing spline in (4) there is no penalty term on the functions of the categorical predictors (h_j).

Generalizing to the ACOSSO Estimate. The COmponent Selection and Shrinkage Operator (COSSO) [15] penalizes on the sum of the semi-norms instead of the squared semi-norms as in Eq. (4). A semi-norm is a norm which can assign zero to some nonzero elements of the space. In this case, all functions with zero second derivative (i.e., linear functions) will have zero penalty (i.e., semi-norm equal to zero). For ease of presentation, we will continue to restrict attention to the additive model. However, all of the following discussion applies directly to the two-way (or higher) interaction model as well.

The additive COSSO estimate, $\hat{f}(\mathbf{w}) = \sum \hat{g}_i(x_i) + \sum \hat{h}_j(z_j)$, is given by the function $f \in \mathcal{F}_{add}$ that minimizes

$$\frac{1}{N} \sum_{n=1}^{N} [y_n - f(\mathbf{w}_n)]^2 + \lambda_1 \sum_{i=1}^{I} \left\{ \left[\int_0^1 g_i'(x_i)dx_i \right]^2 + \int_0^1 [g_i''(x_i)]^2 \, dx_i \right\}^{1/2} +$$

$$\lambda_2 \sum_{j=1}^{J} \left\{ \sum_{z_j=1}^{b_j} h_j^2(z_j) \right\}^{1/2}. \tag{5}$$

There are four key differences in the penalty term in Eq. (5) relative to the additive smoothing spline of Eq. (4). First, there is an additional term $\left[\int_0^1 g_i'(x_i)dx_i \right]^2$ in the penalty for continuous predictor functional components, which can also be written $[g_i(1) - g_i(0)]^2$, that penalizes the magnitude of the overall trend of the functional components g_i that correspond to continuous predictors. Second, there is now a penalty on the $\mathcal{L}^2$ norm of the h_j that correspond to the categorical predictors. Third, in contrast to the squared semi-norm in the additive smoothing spline, each term in the sum in the penalty in Eq. (5) can be thought of as a semi-norm over functions $g_i \in \mathcal{G}_i$ or $h_j \in \mathcal{H}_j$, respectively, (only constant functions have zero penalty). This encourages some of the terms in the sum to be exactly zero. Fourth, the COSSO penalty only has two tuning parameters (three if two-way interactions are included), which can be chosen via GCV or similar means.

Finally, ACOSSO is a weighted version of COSSO, where a rescaled semi-norm is used as the penalty for each of the functional components. Specifically, we select as our estimate the function $f \in \mathcal{F}_{add}$ that minimizes

$$\frac{1}{N} \sum_{n=1}^{N} [y_n - f(\mathbf{w}_n)]^2 + \lambda_1 \sum_{i=1}^{I} v_i \left\{ \left[\int_0^1 g_i'(x_i)dx_i \right]^2 + \int_0^1 [g_i''(x_i)]^2 \, dx_i \right\}^{1/2} +$$

$$\lambda_2 \sum_{j=1}^{J} w_j \left\{ \sum_{z_j=1}^{b_j} h_j^2(z_j) \right\}^{1/2}, \tag{6}$$

where the v_i, w_j, $0 < v_i, w_j \leq \infty$, are weights that can depend on an initial estimate of f which we denote $\tilde{f}$. Our implementation of ACOSSO takes $\tilde{f}$ to be the COSSO estimate of Eq. (5), in which λ_1 and λ_2 are chosen by the GCV criterion. We then use

$$v_i = \left[\int_0^1 \tilde{g}_i^2(x_i)dx_i \right]^{-1} \quad \text{for } i = 1, \dots, I$$

$$w_j = \left(\frac{1}{b_j} \sum_{z_j=1}^{b_j} \tilde{h}_j^2(z_j) \right)^{-1} \quad \text{for } j = 1, \dots, J. \tag{7}$$

This allows for more flexible estimation (less penalty) on the functional components that show more signal in the initial estimate. As shown in [29], this approach results in better performance on many test cases and more favorable asymptotic properties than COSSO.

The minimizer of the expression in Eq. (6) is obtained using an iterative algorithm and a RKHS framework similar to that used to find the minimizer of Eq. (4), see [29] for more details on the computation of the solution. The two-way interaction model is used in the results of Section 4.

2.2 Gaussian Processes for Models with Quantitative and Qualitative Factors

This section describes a computationally efficient method developed in Zhou, Qian, and Zhou [39] for fitting Gaussian process models with quantitative and qualitative factors proposed in Qian, Wu, and Wu [20]. Consider a computer model with inputs $\mathbf{w} = (\mathbf{x}^t, \mathbf{z}^t)^t$, where $\mathbf{x} = (x_1, \ldots, x_I)^t$ consists of all the quantitative factors and $\mathbf{z} = (z_1, \ldots, z_J)^t$ consists of all the qualitative factors with z_j having b_j levels. The number of the qualitative levels of $\mathbf{z}$ is given by

$$m = \prod_{j=1}^{J} b_j. \tag{8}$$

Throughout, the factors in $\mathbf{z}$ are assumed to be qualitative but not ordinal. Gaussian process models with ordinal qualitative factors can be found in Section 4.4 of [20]. The response of the computer model at an input value $\mathbf{w}$ is modeled as

$$y(\mathbf{w}) = \mathbf{f}^t(\mathbf{w})\boldsymbol{\beta} + \epsilon(\mathbf{w}), \tag{9}$$

where $\mathbf{f}(\mathbf{w}) = [f_1(\mathbf{w}), \ldots, f_p(\mathbf{w})]^t$ is a set of p user-specified regression functions, $\boldsymbol{\beta} = (\beta_1, \ldots, \beta_p)^t$ is a vector of unknown coefficients and the residual $\epsilon(\mathbf{w})$ is a stationary Gaussian process with mean 0 and variance σ^2. The model in (9) has a more general form than the standard Gaussian process model with only quantitative factors $\mathbf{x}$ given by

$$y(\mathbf{x}) = \mathbf{f}^t(\mathbf{x})\boldsymbol{\beta} + \epsilon(\mathbf{x}), \tag{10}$$

where $\mathbf{f}(\mathbf{x}) = [f_1(\mathbf{x}), \ldots, f_p(\mathbf{x})]^t$ is a set of p user-specified regression functions depending on $\mathbf{x}$ only, $\boldsymbol{\beta} = (\beta_1, \ldots, \beta_p)^t$ is a vector of unknown coefficients, and the residual $\epsilon(\mathbf{x})$ is a stationary Gaussian process with mean 0, variance σ^2 and a correlation function for $\mathbf{x}$.

For m in (8), let $c_1, \ldots, c_m$ denote the m qualitative levels of $\mathbf{z}$ and let $\mathbf{w} = (\mathbf{x}^t, c_q)^t$ $(q = 1, \ldots, m)$ denote any input value. For two input values $\mathbf{w}_1 = (\mathbf{x}_1^t, c_1)^t$ and $\mathbf{w}_2 = (\mathbf{x}_2^t, c_2)^t$, the correlation between $y(\mathbf{w}_1)$ and $y(\mathbf{w}_2)$ is defined to be

$$\text{cor}\left[\epsilon(\mathbf{w}_1), \epsilon(\mathbf{w}_2)\right] = \tau_{c_1, c_2} \varphi(\mathbf{x}_1, \mathbf{x}_2), \tag{11}$$

where φ is the correlation function for the quantitative factors $\mathbf{x}$ in the model (9) and τ_{c_1,c_2} is the cross-correlation between the qualitative levels c_1 and c_2. The choice of φ is flexible. We use a *Gaussian correlation function* [25]

$$\varphi(\mathbf{x}_1, \mathbf{x}_2) = \exp\left\{ -\sum_{i=1}^{I} \phi_i (x_{1i} - x_{2i})^2 \right\} \tag{12}$$

but other correlation functions such as Wendland's *compactly supported correlation function* [38] may also be used.

The unknown roughness parameters ϕ_i in (12) will be collectively denoted as $\boldsymbol{\Phi} = \{\phi_i\}$. The $m \times m$ matrix $\mathbf{T} = \{\tau_{r,s}\}$, with entries being the cross-correlations among the qualitative levels, must be *positive definite with unit diagonal elements* in order for (11) to be a valid correlation function. This condition can be achieved in two ways. One way is to use the semi-definite programming techniques with positive definiteness constraints [20], which are computationally intensive. [39] provides a more efficient way for modeling $\mathbf{T}$ by using the hypersphere decomposition, originally introduced for modeling correlations in financial applications [21]. This method first applies a Cholesky-type decomposition to $\mathbf{T}$

$$\mathbf{T} = \mathbf{L}\mathbf{L}^t, \tag{13}$$

where $\mathbf{L} = \{l_{r,s}\}$ is a lower triangular matrix with strictly positive diagonal entries. Then, let $l_{1,1} = 1$ and for $r = 2, \ldots, m$, consider a *spherical coordinate system*

$$\begin{cases} l_{r,1} = \cos(\theta_{r,1}), \\ l_{r,s} = \sin(\theta_{r,1}) \cdots \sin(\theta_{r,s-1}) \cos(\theta_{r,s}), \text{ for } s = 2, \ldots, r-1, \\ l_{r,r} = \sin(\theta_{r,1}) \cdots \sin(\theta_{r,r-2}) \sin(\theta_{r,r-1}), \end{cases} \tag{14}$$

where $\theta_{r,s} \in (0, \pi)$. Denote by $\boldsymbol{\Theta}$ all $\theta_{r,s}$ involved in (14).

Suppose that the computer model under consideration is conducted at n different input values, $D_{\mathbf{w}} = (\mathbf{w}_1^0, \ldots, \mathbf{w}_n^0)$, with the corresponding response values denoted by $\mathbf{y} = (y_1, \ldots, y_n)^t$. The parameters in model (9) to be estimated are σ^2, β, $\boldsymbol{\Phi}$ and $\boldsymbol{\Theta}$. The *maximum likelihood estimators* of these parameters are denoted by $\hat{\sigma}^2$, $\hat{\beta}$, $\hat{\boldsymbol{\Phi}}$ and $\hat{\boldsymbol{\Theta}}$, respectively. The log-likelihood function of $\mathbf{y}$, up to an additive constant, is

$$-\frac{1}{2}\left[n\log(\sigma^2) + \log(|\mathbf{R}|) + (\mathbf{y} - \mathbf{F}\beta)^t \mathbf{R}^{-1}(\mathbf{y} - \mathbf{F}\beta)/\sigma^2 \right], \tag{15}$$

where $\mathbf{F} = \left[\mathbf{f}(\mathbf{w}_1^0), \ldots, \mathbf{f}(\mathbf{w}_n^0)\right]^t$ is an $n \times p$ matrix and $\mathbf{R}$ is the correlation matrix with (i,j)th entry cor $\left[\epsilon(\mathbf{w}_i^0), \epsilon(\mathbf{w}_j^0)\right]$ defined in (11). Given $\boldsymbol{\Phi}$ and $\boldsymbol{\Theta}$, $\hat{\beta}$ and $\hat{\sigma}^2$ are

$$\hat{\beta} = (\mathbf{F}^t \mathbf{R}^{-1} \mathbf{F})^{-1} \mathbf{F}^t \mathbf{R}^{-1} \mathbf{y}, \tag{16}$$

and

$$\hat{\sigma}^2 = (\mathbf{y} - \mathbf{F}\hat{\beta})^t \mathbf{R}^{-1}(\mathbf{y} - \mathbf{F}\hat{\beta})/n. \tag{17}$$

Plugging (16) and (17) into (15), $\hat{\boldsymbol{\Phi}}$ and $\hat{\boldsymbol{\Theta}}$ can be obtained as

$$(\hat{\boldsymbol{\Phi}}, \hat{\boldsymbol{\Theta}}) = \arg\min_{\boldsymbol{\Phi}, \boldsymbol{\Theta}}\{n\log(\hat{\sigma}^2) + \log(|\mathbf{R}|)\}. \tag{18}$$

The optimization problem in (18) only involves the constraints that $\theta_{r,s} \in (0, \pi)$ for $\hat{\boldsymbol{\Theta}}$ and $\phi_i \geq 0$ for $\hat{\boldsymbol{\Phi}}$. It can be solved by modifying the DACE toolbox in Matlab [16] to incorporate the reparameterization in (14). A small nugget term is added to the diagonals of $\mathbf{R}$ to mitigate potential singularity. The fitted model can be used to predict the response value y at any untried input value. Given $\hat{\sigma}^2$, $\hat{\boldsymbol{\beta}}$, $\hat{\boldsymbol{\Phi}}$ and $\hat{\boldsymbol{\Theta}}$, the *empirical best linear unbiased predictor* (EBLUP) of y at any input value $\mathbf{w}_0$ is

$$\hat{y}(\mathbf{w}_0) = \mathbf{f}^t(\mathbf{w}_0)\hat{\boldsymbol{\beta}} + \hat{\mathbf{r}}_0^t \hat{\mathbf{R}}^{-1}(\mathbf{y} - \mathbf{F}\hat{\boldsymbol{\beta}}), \tag{19}$$

where $\hat{\mathbf{r}}_0 = \left\{\text{cor}\left[\epsilon(\mathbf{w}_0^0), \epsilon(\mathbf{w}_1^0)\right], \ldots, \text{cor}\left[\epsilon(\mathbf{w}_0^0), \epsilon(\mathbf{w}_n^0)\right]\right\}^t$ and $\hat{\mathbf{R}}$ is the estimated correlation matrix of $\mathbf{y}$. Similarly to its counterpart for the Gaussian process model in (10) with quantitative factors, the EBLUP in (19) smoothly interpolates all the observed data points. Features of the function $\hat{y}(\mathbf{w})$ can be visualized by plotting the estimated functional main effects and interactions.

In this work, we consider four methods for building Gaussian process models for a computer experiment with qualitative and quantitative factors.

- The individual Kriging method, denoted by IK. This method fits data associated with different qualitative levels separately using distinct Gaussian process models for the quantitative variables in (10).
- The *exchangeable correlation* method for the qualitative factors, denoted by EC. It assumes the cross-correlation $\tau_{r,s}$ in (11) to be

$$\tau_{r,s} = c\,(0 < c < 1) \quad \text{for } r \neq s.$$

- The *multiplicative correlation* method for the qualitative factors, denoted by MC. It assumes the cross-correlation $\tau_{r,s}$ in (11) to be

$$\tau_{r,s} = \exp\{-(\theta_r + \theta_s)I[r \neq s]\}\,(\theta_r, \theta_s > 0).$$

- The method proposed in (13) and (14) with an *unrestricted* correlation function for the qualitative factors, denoted by UC.

2.3 Treed Gaussian Processes (TGP)

In practice, many situations involving the emulation of computer models call for more flexibility than is reasonable under the common assumption of stationarity. However, a fully nonstationary model may be undesirable as well, because of the vastly increased difficulty of performing inference due to a nonstationary model's complexity. A good compromise can be local stationarity. A treed Gaussian process (TGP) [8] is designed to take advantage of local stationarity. It defines a treed partitioning process on the predictor space and fits distinct, but hierarchically related, stationary GPs to separate regions at the leaves. The treed form of the partition makes the model easily interpretable: having the treed partitions with separate GPs makes it easy to identify the GP model in each branch. At the same time, the partitioning results in smaller matrices for inversion than would be required under a standard GP model and thereby provides a nonstationary model that actually facilitates faster inference. Using a fully Bayesian approach allows for model averaging across the tree space, resulting in smooth and continuous fits when the data are not naturally partitioned. The partitions are fit simultaneously with the individual GP parameters using reversible jump Markov chain Monte Carlo, so that all parts of the model can be learned automatically from the data. The posterior predictive distribution thus takes into account uncertainty from the data, from the fitted parameters, and from the fitted partitions.

TGP inherits its partitioning scheme from simpler treed models such as CART [3] and BCART (for Bayesian CART) [5] . Each uses recursive binary splits so that each branch of the tree in any of these models divides the predictor space in two, with multiple splits allowed on the same variable for full flexibility. Consider predictors $x \in R^P$ for some split dimension $p \in \{1, ..., P\}$ and split value v, points with $x_p \leq v$ are assigned to the left branch, and points with $x_p > v$ are assigned to the right branch. Partitioning is recursive and may occur on any input dimension p, so arbitrary axis-aligned regions in the predictor space may be defined. Conditional on a treed partition, models are fit in each of the leaf regions. In CART the underlying models are "constant" in that only the mean and standard deviation of the real-valued outputs are inferred. TGP fits a Gaussian process Z_ν in each leaf ν using the following hierarchical model:

$$\mathbf{Z}_\nu | \boldsymbol{\beta}_\nu, \sigma_\nu^2, \mathbf{K}_\nu \sim N_{n_\nu}(\mathbf{F}_\nu \boldsymbol{\beta}_\nu, \sigma_\nu^2 \mathbf{K}_\nu) \quad \boldsymbol{\beta}_0 \sim N_m(\mu, \mathbf{B})$$

$$\sigma_\nu^2 \sim IG(\alpha_\sigma/2, q_\sigma/2)$$

$$\boldsymbol{\beta}_\nu | \sigma_\nu^2, \tau_\nu^2, \mathbf{W}, \boldsymbol{\beta}_0 \sim N_m(\boldsymbol{\beta}_0, \sigma_\nu^2 \tau_\nu^2 \mathbf{W}) \quad \mathbf{W}^{-1} \sim W((\rho\mathbf{V})^{-1}, \rho)$$

$$\tau_\nu^2 \sim IG(\alpha_\tau/2, q_\tau/2) \tag{20}$$

where $\mathbf{F}_\nu = (\mathbf{1}, \mathbf{X}_\nu)$ contains the data in that leaf. N, IG, and W are the Multivariate Normal, Inverse–Gamma, and Wishart distributions, respectively. $\mathbf{K}_\nu$ is the separable power family covariance matrix with a nugget.

Classical treed methods, such as CART, can cope quite naturally with categorical, binary, and ordinal inputs. For example, categorical inputs can be encoded in binary, and splits can be proposed with rules such as $x_p < 1$. Once a split is made on a binary input, no further process is needed, marginally, in that dimension. Ordinal inputs can also be coded in binary, and thus treated as categorical, or treated as real-valued and handled in a default way. This formulation presents an alternative to that of Section 2.2. While that formulation allows a powerful and flexible representation of qualitative inputs in the model, it does not allow for nonstationarity. TGP allows the combination of qualitative inputs and nonstationary modeling.

Rather than manipulate the GP correlation to handle categorical inputs, the tree presents a more natural mechanism for such binary indicators. That is, they can be included as candidates for treed partitioning but ignored when it comes to fitting the models at the leaves of the tree. The benefits of removing the Booleans from the GP model(s) go beyond producing full-rank design matrices at the leaves of the tree. Loosely speaking, removing the Boolean indicators from the GP part of the treed GP gives a more parsimonious model. The tree is able to capture all of the dependence in the response as a function of the indicator input, and the GP is the appropriate nonlinear model for accounting for the remaining relationship between the real-valued inputs and outputs. Further advantages to this approach include speed (a partitioned model gives smaller covariance matrices to invert) and improved mixing in the Markov chain when a separable covariance function is used. Finally, the treed model allows the practitioner to immediately ascertain whether the response is sensitive to a particular categorical input by tallying the proportion of time the Markov chain visited trees with splits on the corresponding binary indicator. A much more involved Monte Carlo technique (e.g., following [24]) would otherwise be required in the absence of the tree. Here we use the implementation developed by Broderick and Gramacy [4].

3   Testing and Assessment Approach

In order to evaluate and compare the three mixed-variable surrogate modeling approaches, we established a common experimental strategy that can be consistently applied to all of them. There are three primary components to which we paid particular attention. They are the test functions, the sample design used for surrogate construction, and the comparison metrics. Each is described in the following subsections.

3.1   Test Functions

In order to consolidate a common set of test functions on which to evaluate the different surrogate approaches, we developed a generic C++ testbed. It was designed and developed to meet the following requirements:

- ability to control the number of discrete variables and the number of levels per discrete variable in order to test method scalability with respect to these features,
- ability to control problem complexity in order to evaluate performance on a variety of problems,
- extendable in order to easily add new functions, and
- easy to use in multiple computing environments across all surrogate software packages.

The testbed include both defined hard-wired functions and randomly-generated polynomial functions. The latter is based on work by McDaniel and Ankenman [17]. We refer the reader to [33] for more details on the implementation.

3.2 Sample Design

The accuracy of a response surface surrogate can be affected by the number of data points used to build it as well as how those points are chosen. Therefore, we vary the number and design of build points in our numerical experiments. All designs are based on Latin Hypercube designs (LHD) of the parameter space. We define n to be the number of LHD runs per qualitative level of the categorical variables and m to be the number of discrete levels (or combinations of levels). The total number of points used to build each surrogate is mn. We consider $n = 10, 20, 40, 80$, and the sample design for each training set is constructed in three different ways.

Standard Latin Hypercube. In this approach, one Latin Hypercube design of size mn is generated over all of the continuous parameters. It is then randomly split it into m groups of n samples, and each group is assigned a qualitative level of the categorical variables.

k Latin Hypercube. In this approach, a separate Latin Hypercube design is generated for every given level of categorical variables. That is, we generate m independent Latin hypercube designs, each of size n and corresponding to one qualitative level.

Sliced Latin Hypercube. This approach is based on recent work by Qian [19]. This design is a Latin Hypercube for the continuous factors and is sliced into groups of smaller Latin Hypercube designs associated with different categorical levels. In this case, we generate a sliced Latin hypercube design with m slices, where each slice of n runs corresponds to one qualitative level.

Because of the randomness associated with the LHS samples, we generate 10 replicate training sets for each combination of n and design type.

3.3 Comparison Metric

Evaluating the performance of computational methods can be challenging, particularly with regard to the accuracy of the method. This is because the accuracy required for different applications of the method can vary. In this study, our primary focus is on gaining an understanding of the accuracy of mixed variable surrogates relative to each other, so we use a relatively fine-grained metric. In particular, we use mean squared error between surrogate predictions and true function values over a set of given points. For every replication of a given n and training design type, the mean squared errors (MSE) are calculated based on a testing set using a Latin hypercube design with 200 samples for each qualitative level. We then compare the mean and spread of the errors. Lower values of these quantities constitute better performance.

4 Results

We present selected results of our evaluation of the surrogate approaches discussed in Section 2. Specifically, we compared the results of TGP, ACOSSO, and the Gaussian process model with the various special correlation functions. Generally, we found that sample design type (standard Latin Hypercube, kLHD, or sliced LHD) did not have a large effect on the MSE. Thus, we show only the results using the sliced LHD. We applied these different surrogate methods to a set of test functions to be described and compared the results over different training set sizes.

4.1 Test Function 2

The first function we considered in our numerical experiments has one categorical variable with five levels. It also two continuous variables, both of which fall between the values of 0 and 1. This function has regions where the responses at the different categorical levels are very similar. This will allow us to evaluate how well the different surrogate approaches can resolve the different discrete levels.

$$f(x) = \begin{cases} \sin(2\pi x_3 - \pi) + 7\sin^2(2\pi x_2 - \pi) & \text{if } x_1 = 1 \\ \sin(2\pi x_3 - \pi) + 7\sin^2(2\pi x_2 - \pi) + 12.0\sin(2\pi x_3 - \pi) & \text{if } x_1 = 2 \\ \sin(2\pi x_3 - \pi) + 7\sin^2(2\pi x_2 - \pi) + 0.5\sin(2\pi x_3 - \pi) & \text{if } x_1 = 3 \\ \sin(2\pi x_3 - \pi) + 7\sin^2(2\pi x_2 - \pi) + 8.0\sin(2\pi x_3 - \pi) & \text{if } x_1 = 4 \\ \sin(2\pi x_3 - \pi) + 7\sin^2(2\pi x_2 - \pi) + 3.5\sin(2\pi x_3 - \pi) & \text{if } x_1 = 5 \end{cases}$$

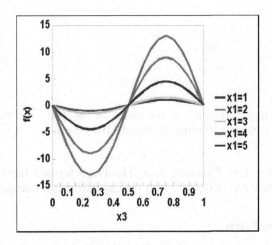

Fig. 1. Test Function 2

Figures 2-3 give the boxplots of the MSEs of the four methods for $n = 10, 20$. The Y axis is the mean squared error (MSE) of the surrogate construction. The Gaussian correlation function in (12) for the quantitative factors is used in the IK, EC, MC and UC methods. Note that the Y-axis scale is different on these figures. Ideally, it would be nice to see the MSE plotted on the same scale so that it is easy to see the decrease in error as a function of the number of training samples. However, the MSE varied so dramatically for some of these results that keeping an MSE scale to allow for plotting maximum MSE values would result in the reader not seeing the differences in situations where the MSE was low.

Overall, ACOSSO does very well on this function and outperforms the other methods, especially at the smaller sample levels of $n = 10$ and $n = 20$. We expect this is because the structure of the ACOSSO surrogate maps naturally to separable functions such as this one. For the four GP correlation schemes, the EC, MC and UC methods outperform the IK method.

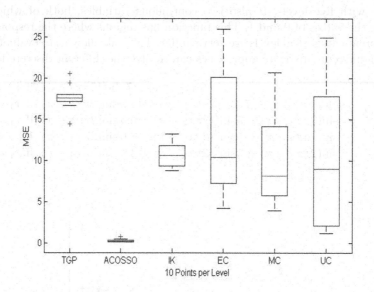

Fig. 2. Test Function 2. Boxplots of the MSEs for the TGP, ACOSSO, IK, EC, MC and UC methods with $n = 10$ using the sliced LHD scheme.

In summary for Test Function 2: ACOSSO performed the best overall, the GP variations with IK, EC, MC and UC methods also performed well.

4.2   Goldstein-Price

The second function we considered is the Goldstein-Price function. It has one continuous variable and one discrete variable. The discrete variable, x_1, can take on the values of $-2, 0$, and 2. The continuous variable, x_2, ranges between the values of -2 and 2. This function varies by five orders of magnitude over the domain we chose, so we performed the surrogate construction in log space and the error is presented in log space.

$$f(x) = (1 + (x_1 + x_2 + 1)^2 \cdot (19 - 14x_1 + 3x_1^2 - 14x_2 + 6x_1x_2 + 3x_2^2)) \cdot$$
$$(30 + (2x_1 - 3x_2)^2 \cdot (18 - 32x_1 + 12x_1^2 + 48x_2 - 36x_1x_2 + 27x_2^2))$$

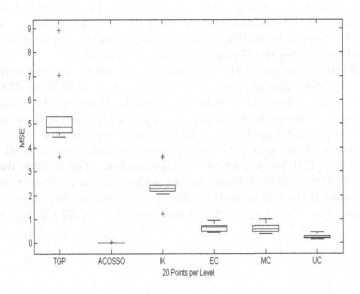

Fig. 3. Boxplots of the MSEs for the TGP, ACOSSO, IK, EC, MC and UC methods with $n = 20$ using the sliced LHD scheme

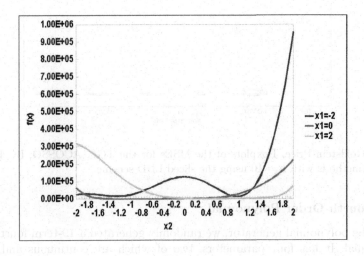

Fig. 4. Goldstein Price Function

Figures 5-6 give the boxplots of the MSEs of the four methods for $n = 10, 20, 40$. In these figures, the Y axis is the mean squared error of the surrogate in log space. For the Gaussian process model, the four variations of IK, EC, MC and UC methods all used the compact support Wendland correlation function. For the Goldstein-Price function, the compactly supported correlation performed better than the Gaussian correlation function.

Overall, the variations of the Gaussian process model do very well on this function. ACOSSO also performs well, and the mean MSE from ACOSSO is close to the mean from the various GP methods. However, the variability of the ACOSSO results is slightly larger, as shown in Figures 5-6. Note that TGP has larger MSE at all sample levels. However, when we performed the surrogate construction in the original space without taking the logarithm of the Goldstein-Price function, TGP outperformed the other methods. This is likely due to the ability of TGP to identify different regions of the space with different properties (e.g. the scale of the Goldstein-Price function is much smaller in the center of the domain than at the edges of the domain we are using for this case study).

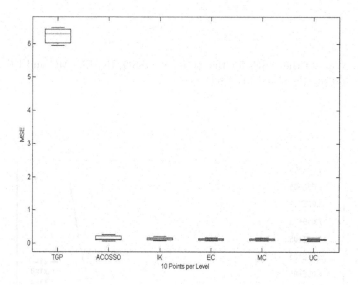

Fig. 5. Goldstein-Price. Boxplots of the MSEs for the TGP, ACOSSO, IK, EC, MC and UC methods with $n = 10$ using the sliced LHD scheme.

4.3   Fourth Order Polynomial

Using the polynomial generator, we randomly generated a 19-term fourth order polynomial. It has four parameters, two of which are continuous and two of which are discrete. The x_3 and x_4 are continuous variables that fall between 0 and 100, and x_1 and x_2 are discrete variables that have three levels, namely 20, 50, and 80. The polynomial is given by the following:

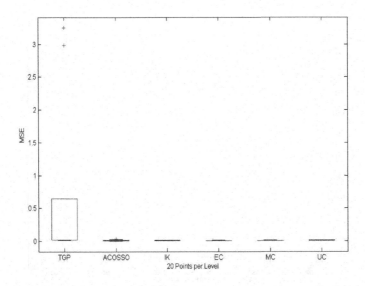

Fig. 6. Goldstein-Price. Boxplots of the MSEs for the TGP, ACOSSO, IK, EC, MC and UC methods with $n = 20$ using the sliced LHD scheme.

$$f(x) = 53.3108 + 0.184901x_1 - 5.02914 \cdot 10^{-6}x_1^3 + 7.72522 \cdot 10^{-8}x_1^4 -$$
$$0.0870775x_2 - 0.106959x_3 + 7.98772 \cdot 10^{-6}x_3^3 + 0.00242482x_4 +$$
$$1.32851 \cdot 10^{-6}x_4^3 - 0.00146393x_1x_2 - 0.00301588x_1x_3 -$$
$$0.00272291x_1x_4 + 0.0017004x_2x_3 + 0.0038428x_2x_4 - 0.000198969x_3x_4 +$$
$$1.86025 \cdot 10^{-5}x_1x_2x_3 - 1.88719 \cdot 10^{-6}x_1x_2x_4 + 2.50923 \cdot 10^{-5}x_1x_3x_4 -$$
$$5.62199 \cdot 10^{-5}x_2x_3x_4$$

The results for the fourth order polynomial are shown in Figures 7-8. These figures show that the Gaussian processes with the various correlation functions such as EC, MC, etc. perform well. Interestingly, ACOSSO does not seem to improve, even as the size of training set increases. That is, the average MSE for ACOSSO with $n = 10$ is 1.5, while the average MSE for ACOSSO with $n = 80$ is 1.4. In contrast, the other approaches all improve the MSE by eight orders of magnitude. We hypothesize that ACOSSO is struggling when there is significant interaction between variables. In particular, it is trying to construct its response as the aggregation of separable functions which may not capture the interactions well.

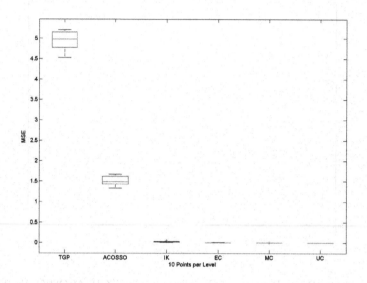

Fig. 7. Fourth-order Polynomial. Boxplots of the MSEs for the TGP, ACOSSO, IK, EC, MC and UC methods with $n = 10$ using the sliced LHD scheme.

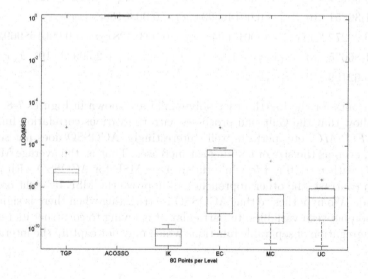

Fig. 8. Fourth-order Polynomial. Boxplots of the MSEs for the TGP, ACOSSO, IK, EC, MC and UC methods with $n = 80$ using the sliced LHD scheme.

5   Summary

This paper investigated three main classes of surrogate methods which can handle "mixed" discrete and continuous variables: adaptive smoothing splines, Gaussian processes with special correlation functions, and Treed Gaussian processes. We chose test problems which were challenging but tractable for repeated comparison runs. The results presented are representative of the extensive comparisons we performed, varying the number of build points used in the surrogate construction, varying the sample designs used, and building multiple surrogates of a given type so that we could compute statistics of the response to give fair comparisons (e.g. so we would not be misled by constructing only one surrogate on one set of build points).

Overall, all methods appear viable for small numbers of categorical variables with a few levels, and we were able to gain some general insights across the wide range of studies performed. ACOSSO and the Gaussian processes with special correlation functions generally performed well. ACOSSO performed best for separable functions, especially at small training set sizes. This is a particularly valuable trait, as computational expense usually prevents large training set sizes. When there are significant interactions between discrete and continuous parameters, as in the fourth-order polynomial, ACOSSO performs poorly even with a larger number of training points (40 or 80). Both results are expected because ACOSSO is constructed over separable functions, and its performance may degrade somewhat when significant interactions between variables are present. The GP with special correlation functions appears the most consistent of all the methods. However, that approach was the most sensitive to build design and did not perform as well with a plain LHD design, whereas ACOSSO and TGP were not significantly affected by the design. TGP success depends on being able to identify splits where individual GPs work well in separate parts of the domain. TGP performs well on poorly scaled functions, but we found it does not perform well when the continuous variables are not predictive for certain combinations of categorical variable levels. These insights will allow us to move forward with applying these surrogate methods to computational analysis problems for which they are best suited.

Acknowledgements. The authors would like to thank a number of colleagues for many fruitful discussions on this topic: William Hart, John Siirola, Jean-Paul Watson, Genetha Gray, Cynthia Phillips, Ali Pinar, and David Woodruff. We also thank Sandia management, specifically M. Daniel Rintoul, and the Laboratory Directed Research and Development office at Sandia for supporting this project. Finally, we would like to thank Ken Perano for writing the C++ software that encodes the test functions we used to evaluate the surrogate approaches.

References

1. Adams, B.M., Bohnhoff, W.J., Dalbey, K.R., Eddy, J.P., Eldred, M.S., Gay, D.M., Hough, P.D., Swiler, L.P.: DAKOTA, a multilevel parallel object-oriented framework for design optimization, parameter estimation, uncertainty quantification, and sensitivity analysis: Version 5.1 user's manual. Technical Report SAND2010-2183, Sandia National Laboratories, Albuquerque, NM (2010), http://dakota.sandia.gov/documentation.html
2. Berlinet, A., Thomas-Agnan, C.: Reproducing Kernel Hilbert Spaces in Probability and Statistics. Kluwer Academic Publishers, Norwell (2004)
3. Breiman, L., Friedman, J.H., Olshen, R., Stone, C.: Classification and Regression Trees. Wadsworth, Belmont (1984)
4. Broderick, T., Gramacy, R.B.: Classification and categorical inputs with treed Gaussian process models. Journal of Classification 28(2), 244–270 (2011)
5. Chipman, H., George, E., McCulloch, R.: Bayesian CART Model Search (with discussion). Journal of the American Statistical Association 93, 930–960
6. Craven, P., Wahba, G.: Smoothing noisy data with spline functions: estimating the correct degree of smoothing by the method of generalized cross-validation. Numerical Mathematics 31, 377–403 (1979)
7. Eubank, R.L.: Nonparametric Regression and Spline Smoothing. CRC Press, London (1999)
8. Gramacy, R.B., Lee, H.K.H.: Bayesian treed gaussian process models with an application to computer modeling. Journal of the American Statistical Association 103, 1119–1130 (2008)
9. Gramacy, R.B., Lee, H.K.H.: Gaussian processes and limiting linear models. Computational Statistics and Data Analysis 53, 123–136 (2008)
10. Gramacy, R.B., Taddy, M.: Categorical inputs, sensitivity analysis, optimization and importance tempering with tgp version 2, an r package for treed gaussian process models. Technical report, R manual (2009), http://cran.r-project.org
11. Gu, C.: Smoothing Spline ANOVA Models. Springer, New York (2002)
12. Hastie, T., Tibshirani, R.J.: Generalized Additive Models. Chapman & Hall/CRC, London (1990)
13. Kennedy, M.C., O'Hagan, A.: Bayesian calibration of computer models (with discussion). Journal of the Royal Statistical Society B 63, 425–464 (2001)
14. Kohavi, R.: A study of cross-validation and bootstrap for accuracy estimation and model selection. In: Proceedings of the Fourteenth International Joint Conference on Artificial Intelligence, vol. 2(12), pp. 1137–1143 (1995)
15. Lin, Y., Zhang, H.: Component selection and smoothing in smoothing spline analysis of variance models. Annals of Statistics 34(5), 2272–2297 (2006)
16. Lophaven, S.N., Neilson, H.B., Sondergaard, J.: Dace - a matlab kriging toolbox. Technical report (2009), http://www2.imm.dtu.dk/~hbn/dace/
17. McDaniel, W.R., Ankenman, B.E.: A response surface test bed. Quality and Reliability Engineering International 16, 363–372 (2000)
18. Neter, J., Wasserman, W., Kutner, M.H.: Applied Linear Statistical Models, 2nd edn. Irwin (1985)
19. Qian, P.Z.G.: Sliced latin hypercube designs (2011) (submitted)

20. Qian, P.Z.G., Wu, H., Wu, C.F.J.: Gaussian process models for computer experiments with qualitative and quantitative factors. Technometrics 50(3), 383–396 (2008)
21. Rebonato, R., Jackel, P.: The most general methodology for creating a valid correlation matrix for risk management and option pricing purposes. The Journal of Risk 2, 17–27 (1999)
22. Reich, B.J., Storlie, C.B., Bondell, H.D.: Variable selection in bayesian smoothing spline anova models: Application to deterministic computer codes. Technometrics 51(2), 110–120 (2009)
23. Sacks, J., Welch, W.J., Mitchel, T.J., Wynn, H.P.: Design and analysis of computer experiments. Statistical Science 4(4), 409–435 (1989)
24. Saltelli, A., Ratto, M., Andres, T., Campolongo, F., Cariboni, J., Gatelli, D., Saisana, M., Tarantola, S.: Global sensitivity analysis, The Primer. Wiley and Sons (2008)
25. Santner, T., Williams, B., Notz, W.: The Design and Analysis of Computer Experiments. Springer, New York (2003)
26. Schimek, M. (ed.): Smoothing and Regression: Approaches, Computation, and Application. John Wiley, New York (2000)
27. Simpson, T.W., Toropov, V., Balabanov, V., Viana, F.A.C.: Design and analysis of computer experiments in multidisciplinary design optimization: A review of how far we have come – or not. In: Proceedings of the 12th AIAA/ISSMO Multidisciplinary Analysis and Optimization Conference, Victoria, British Columbia, Canada (September 2008) AIAA Paper 2008-5802
28. Storlie, C.B., Bondell, H.D., Reich, B.J.: A locally adaptive penalty for estimation of functions with varying roughness. Journal of Computational and Graphical Statistics 19(3), 569–589 (2010)
29. Storlie, C.B., Bondell, H.D., Reich, B.J., Zhang, H.H.: Surface estimation, variable selection, and the nonparametric oracle property. Statistica Sinica 21(2), 679–705 (2010)
30. Storlie, C.B., Helton, J.C.: Multiple predictor smoothing methods for sensitivity analysis: Description of techniques. Reliabililty Engineering and System Safety 93(1), 28–54 (2008)
31. Storlie, C.B., Swiler, L.P., Helton, J.C., Sallaberry, C.J.: Implementation and evaluation of nonparametric regression procedures for sensitivity analysis of computationally demanding models. Reliability Engineering and System Safety 94(11), 1735–1763 (2009)
32. Swiler, L.P., Wyss, G.D.: A user's guide to Sandia's latin hypercube sampling software: LHS UNIX library and standalone version. Technical Report SAND04-2439, Sandia National Laboratories, Albuquerque, NM (July 2004)
33. Swiler, L.P., Hough, P.D., Qian, P., Xu, X., Storlie, C.B., Lee, H.: Surrogate models for mixed discrete-continuous variables. Technical Report SAND2012-0491, Sandia National Laboratories, Albuquerque, NM (August 2012)
34. Tibshirani, R.J.: Regression shrinkage and selection via the lasso. Journal of the Royal Statistical Society B 58, 267–288 (1996)
35. Viana, F.A.C., Haftka, R.T., Steffan Jr., V.: Multiple surrogates: How cross-validation errors can help us obtain the best predictor. Structural and Multidisciplinary Optimization 39(4), 439–457 (2009)

36. Viana, F.A.C., Haftka, R.T., Steffan Jr., V., Butkewitsch, S., Leal, M.F.: Ensemble of surrogates: a framework based on minimization of the mean integrated square error. In: Proceedings of the 49th AIAA/ASME/ASCE/AHS/ASC Structures, Structural Dynamics, and Materials Conference, Schaumburg, IL (April 2008) AIAA Paper 2008–1885
37. Wahba, G.: Spline Models for Observational Data. CBMS-NSF Regional Conference Series in Applied Mathematics (1990)
38. Wu, Z.: Multivariate compactly supported positive definite radial functions. Advances in Computational Mathematics 4, 283–292 (1995)
39. Zhou, Q., Qian, P.Z.G., Zhou, S.: A simple approach to emulation for computer models with qualitative and quantitative factors. Technometrics 53, 266–273 (2011)

Why Ellipsoid Constraints, Ellipsoid Clusters, and Riemannian Space-Time: Dvoretzky's Theorem Revisited

Karen Villaverde[1], Olga Kosheleva[2], and Martine Ceberio[2]

[1] Department of Computer Science, New Mexico State University,
Las Cruces, NM 88003, USA
kvillave@cs.nmsu.edu
[2] University of Texas at El Paso, El Paso, TX 79968, USA
{olgak,mceberio}@utep.edu

Abstract. In many practical applications, we encounter ellipsoid constraints, ellipsoid-shaped clusters, etc. A usual justification for this ellipsoid shape comes from the fact that many real-life quantities are normally distributed, and for a multi-variate normal distribution, a natural confidence set (containing the vast majority of the objects) is an ellipsoid. However, ellipsoids appear more frequently than normal distributions (which occur in about half of the cases). In this paper, we provide a new justification for ellipsoids based on a known mathematical result – Dvoretzky's Theorem.

Keywords: ellipsoids, constraints, clusters, tensors, space-time physics, Dvoretzky's theorem.

1 Formulation of the Problem

Ellipsoids are ubiquitous. In many practical applications, we encounter ellipsoid constraints, ellipsoid-shaped clusters, etc. (see, e.g., [2]), i.e., sets in an n-dimensional space described by the formula

$$\sum_{i=1}^{n}\sum_{j=1}^{n} a_{ij} \cdot x_i \cdot x_j + \sum_{i=1}^{n} a_i \cdot x_i \leq a_0. \qquad (1)$$

Reformulation in terms of tensors. The above formula (1) shows that to describe an ellipsoid, we need to have a vector (= tensor of order 1) a_i and a tensor a_{ij} of order 2.

A usual probabilistic explanation of the ellipsoid shape. A usual justification for this ellipsoid shape comes from the fact that many real-life quantities are normally distributed, and for a multi-variate normal distribution, a natural confidence set (containing the vast majority of the objects) is an ellipsoid.

M. Ceberio and V. Kreinovich (eds.), *Constraint Programming and Decision Making*,   203
Studies in Computational Intelligence 539,
DOI: 10.1007/978-3-319-04280-0_22, © Springer International Publishing Switzerland 2014

Indeed, it is known that uncertainty can be often described by the Gaussian (= normal) distribution, with the probability density

$$\rho(x) = \frac{1}{\sqrt{2\pi}} \cdot \exp\left(-\frac{(x-a)^2}{2\sigma^2}\right). \tag{2}$$

This possibility comes from the Central Limit Theorem (see, e.g., [12]), according to which the sum $x = \sum_{i=1}^{N} x_i$ of a large number N of independent small random variables x_i has an approximately Gaussian distribution. (To be more precise, the theorem says that in the limit $N \to \infty$, the distribution of the sum tends to the Gaussian distribution.)

In practice, often, the measurement error is caused by a joint effect of a large number of small independent factors, so it makes sense to conclude that the distribution is approximately Gaussian. This theoretical conclusion has been experimentally confirmed on the example of many actual measuring instruments; see, e.g., [9].

The above result is about the 1-D distribution: of a random number. For the multi-D case – of a random vector $x = (x_1, \ldots, x_n)$ – a similar result also leads to multi-D Gaussian distribution, with an expression

$$\rho(x) = \text{const} \cdot \exp\left(-\sum_{i=1}^{n}\sum_{j=1}^{n} c_{ij} \cdot (x_i - a_i) \cdot (x_j - a_j)\right). \tag{3}$$

This probability density $\rho(x)$ is everywhere positive; thus, in principle, an arbitrary tuple Δx is possible. In practical statistics, however, tuples with very low probability density $\rho(\Delta x)$ are considered impossible. For example, in 1-dimensional case, we have a "three sigma" rule: values for which $|\Delta x_1| > 3\sigma_1$ are considered to be almost impossible. In the multi-dimensional case, it is natural to choose some threshold $t > 0$, and consider only tuples for which $\rho(\Delta x) \geq t$ as possible ones. This formula is equivalent to $\ln(\rho(x)) \geq \ln(t)$. For Gaussian distribution, this equality takes the form $\sum_{i=1}^{n}\sum_{j=1}^{n} c_{ij} \cdot (x_i - a_i) \cdot (x_j - a_j) \leq -\ln(t)$, i.e., the form of an ellipsoid.

Problem. While the probabilistic explanation is convincing, it does not cover all the cases. Indeed, according to [9], normal distributions occur in approximately half of the cases, while in many practical applications, ellipsoids appear more frequently.

How can we explain this ubiquity of ellipsoids?

Taylor expansion: a possible explanation. Another possible explanation comes from the fact that the function $g(x_1, \ldots, x_n)$ describing a general constraint $g(x_1, \ldots, x_n) \leq 0$ is usually smooth; thus, it can be usually expanded in Taylor series. In this expansion, terms of higher order become smaller and smaller, so we can usually safely keep only a few first terms in this expansion. In particular, if we only keep linear and quadratic terms, we get an expression (1) – i.e., an ellipsoid.

This argument is reasonable, but it does not explain why in most cases, the first two terms are sufficient and not, e.g., the first three – which would lead to more complex shapes of constraints and clusters (and the use of tensors of higher order).

Comment. An alternative explanation comes from the fact that ellipsoids are known to be the *optimal* approximation sets for different problems with respect to several reasonable optimality criteria; see, e.g., [5, 6]. However, they are optimal only if we consider approximating families of sets characterized by the smallest possible number of parameters.

2 New Explanation Based on Dvoretzky's Theorem

What is Dvoretzky's Theorem. In this paper, we propose a new explanation of the ubiquity of ellipsoids. This explanation is based on a mathematical result called Dvoretzky's theorem.

The original version of this theorem [3] answered a question raised in 1956 by Alexander Grothendieck, one of the most important mathematicians of the 20 century. A. Dvoretzky proved that Grothendieck's hypothesis is indeed true, and that in general, convex sets in large dimensions have sections whose shape is close to ellipsoidal – the larger the dimension, the close this shape to the shape of an ellipsoid.

In 1971, V. L. Milman [7] strengthened this result by proving that not only *there exists* an almost ellipsoidal shape, but also that *almost all* low-dimensional sections of a convex set have an almost ellipsoidal shape. (Strictly speaking, he proved that for every $\varepsilon > 0$, the probability to get a shape which is more than ε-different from ellipsoidal goes to 0 as the dimension of the convex set increases.)

How Dvoretzky's theorem explains the ubiquity of ellipsoid clusters and ellipsoid constraints. In clustering, one of the main problems is that usually, we only measure a few quantities, not enough to easily classify objects. For example, in military applications, the need to classify sonar records into submarine sounds, whale sounds, and noise comes from the fact that we only have a weak (partially observed) signal. Based on a high-quality low-noise recording, it is relatively easy to distinguish between sounds produced by submarines and sounds produced by whales.

Theoretically, each real-life object can be characterized by a point (vector) containing the results of measuring all possible quantities characterizing this object. In this theoretical description, objects are represented by points in a (very) high-dimensional space, and natural classes of objects are sets in this high-dimensional space.

However, in the real world, we only observe a few of these quantities. Thus, what we observe is a lower-dimensional section of a high-dimensional set – and we know that, according to Dvoretzky's theorem, this section is almost always almost ellipsoidal.

A similar argument can be made about constraints. The actual physical constraints depend not only on the observed quantities $x_1, \ldots, x_n$, they also depend on other quantities whose values we do not measure in our experiments. For example, to avoid unnecessary side effects, it is usually recommended that the amount x_1 of a medicine that a doctor prescribes to a patient must lie within bounds depending on the patient's body weight x_2. In other words, we have a constraint of the type $x_1 \leq k \cdot x_2$, where the constant k depends on the specific medicine. However, the actual effect of the medicine depends not only on the body weight, it depends on many other characteristics of a patient – such as physical fitness, general allergic reactions – characteristics that usually, we do not measure. Similarly, in recipes for cooking, the amount of salt x_1 is usually listed depending on the amount of, say, meat x_2 used in the cooking. However, in reality, it should depend also on the parameters that a usual cook does not measure exactly – such as the humidity in the air, etc. (That is why, in contrast to typical US cookbooks that list the exact amounts of all the ingredients, in Mexican, Russian, and French cookbooks these amounts are only approximately listed – so that a skilled cook can take into account other parameters that are difficult to measure :-)

In general, a physical constraint actually has a form $g(x_1, \ldots, x_n, x_{n+1}, \ldots, x_N) \leq 0$, where $x_{n+1}, \ldots, x_N$ are quantities that we do not measure in this particular experiment. Thus, the corresponding n-dimensional constraint set $\{x = (x_1, \ldots, x_n) : g(x_1 \ldots, x_n) \leq 0\}$ is a *section* of the actual (unknown) multi-dimensional constraint set $\{x = (x_1, \ldots, x_n, \ldots, x_N) : g(x_1 \ldots, x_n, \ldots, x_N) \leq 0\}$ – and we already know that in almost all cases, such sections are almost ellipsoidal.

Auxiliary result: why Riemannian space-time? A similar argument can explain why, contrary to physicists' expectations, experiments seem to confirm the Riemannian models of space-time. Before we provide this explanation, let us briefly explain what is the Riemannian model and why physicists expected it to be experimentally disproved.

Before Einstein's General Relativity theory, it was assume that space in Euclidean, i.e., that, in appropriate coordinates, the distance $d(x, x + \Delta x)$ between two close points can be described as $d^2(x, x + \Delta x) = \sum_{i=1}^{n} (\Delta x_i)^2$. In general (not necessarily orthonomal) coordinates, this distance takes a more general form $\sum_{i=1}^{n} \sum_{j=1}^{n} g_{ij} \cdot \Delta x_i \cdot \Delta x_j$. Einstein suggested that the space-time is *locally* Euclidean, so that in the small vicinity of each point, there are coordinates in which the distance is Euclidean – but there are no coordinates in which the distance formula is Euclidean at all the points [8]. Such spaces are known as *Riemannian*.

Einstein himself experimented with extending his theory from the usual (observed) $(3+1)$-dimensional space-time to space-times of higher dimension [4].

It later turned out that higher dimensions are needed to make quantum field theory consistent; see, e.g., [11].

A local Euclidean metric can be characterized by the fact that in this metric, the unit ball is an ellipsoid. In principle, there are other metrics (e.g., l^p-metric for which $d^p(x, x + \Delta x) = \sum_{i=1}^{n} |\Delta x_i|^p$) with different convex bodies for unit balls. The corresponding generalization of Riemannian space-time is called a *Finsler space* [1, 10].

One of the main ideas of quantum physics is that in contrast to classical physics, where, e.g., some trajectories are allowed and some are not, in quantum physics, all trajectories are allowed – just the probability of very non-standard probabilities is small. Similarly, metrics should not be limited to Riemannian metrics, Finsler metrics should also be possible – with some probability. However, while experiments confirm non-standard trajectories of quantum particles and non-standard behavior of quantum fields, surprisingly, all experimental data so far confirms Riemannian metric.

Dvoretzky's theorem explains this phenomenon: indeed, the actual space is multi-dimensional, so we only observe a section of the corresponding convex unit ball, and such a section is close to an ellipsoid.

References

1. Chern, S.-S., Shen, Z.: Riemann-Finsler Geometry. World Scientific, Singapore (2005)
2. Chernousko, F.L.: State Estimation for Dynamic Systems. CRC Press, Boca Raton (1994)
3. Dvoretzky, A.: Some results on convex bodies and Banach spaces. In: Proceedings of the 1960 International Symposium on Linear Spaces, pp. 123–160. Jerusalem Academic Press, Pergamon Press, Jerusalem, Oxford (1961)
4. Einstein, A., Bergmann, P.: On the generalization of Kaluza's theory of electricity. Ann. Phys. 39, 683–701 (1938)
5. Finkelstein, A., Kosheleva, O., Kreinovich, V.: Astrogeometry, error estimation, and other applications of set-valued analysis. ACM SIGNUM Newsletter 31(4), 3–25 (1996)
6. Li, S., Ogura, Y., Kreinovich, V.: Limit Theorems and Applications of Set Valued and Fuzzy Valued Random Variables. Kluwer Academic Publishers, Dordrecht (2002)
7. Milman, V.D.: A new proof of A. Dvoretzky's theorem on cross-sections of convex bodies. Functional Analysis and Its Applications 5(4), 28–37 (1971) (in Russian)
8. Misner, C.W., Thorne, K.S., Wheeler, J.A.: Gravitation. W.H. Freeman, New York (1973)
9. Novitskii, P.V., Zograph, I.A.: Estimating the Measurement Errors. Energoatomizdat, Leningrad (1991) (in Russian)
10. Pavlov, D.G., Atanasiu, G., Balan, V. (eds.): Space-Time Structure. Algebra and Geometry. Russian Hypercomplex Society, Lilia Print, Moscow (2007)
11. Polchinski, J.: String Theory, vols. 1, 2. Cambridge University Press (1998)
12. Rabinovich, S.: Measurement Errors and Uncertainties: Theory and Practice. Springer, New York (2005)

Author Index

Printed in the United States
By Bookmasters